STUDENT IN-CLASS NOTEBOOK

CHARLES A. DANA CENTER
The University of Texas at Austin

MYMATHLAB® FO QUANTITATIVE REASONING

Charles A. Dana Center
The University of Texas at Austin

PEARSON

Boston Columbus Indianapolis New York San Francisco
Amsterdam Cape Town Dubai London Madrid Milan Munich Paris Montreal Toronto
Delhi Mexico City São Paulo Sydney Hong Kong Seoul Singapore Taipei Tokyo

Licensing for State of Texas Community Colleges

General Information About the Dana Center's Copyright

We use all funds generated through use of our materials to further our nonprofit mission. Please send your permission requests or questions to us at this address:

Charles A. Dana Center
The University of Texas at Austin
1616 Guadalupe Street, Suite 3.206
Austin, TX 78701-1222

Fax: 512-232-1855
danaweb@austin.utexas.edu
www.utdanacenter.org

Reproduced by Pearson from electronic files supplied by the author.

ISBN-13: 978-0-13-450721-7
ISBN-10: 0-13-450721-5

6

As always, we welcome your comments and suggestions for improvements. Please contact us at danaweb@austin.utexas.edu or at the mailing address above.

About the Charles A. Dana Center at The University of Texas at Austin

The Dana Center develops and scales math and science education innovations to support educators, administrators, and policy makers in creating seamless transitions throughout the K–14 system for all students, especially those who have historically been underserved.

We work with our nation's education systems to ensure that every student leaves school prepared for success in postsecondary education and the contemporary workplace—and for active participation in our modern democracy. We are committed to ensuring that the accident of where a student attends school does not limit the academic opportunities he or she can pursue. Thus, we advocate for high academic standards, and we collaborate with local partners to build the capacity of education systems to ensure that all students can master the content described in these standards.

Our portfolio of initiatives, grounded in research and two decades of experience, centers on mathematics and science education from prekindergarten through the early years of college. We focus in particular on strategies for improving student engagement, motivation, persistence, and achievement.

We help educators and education organizations adapt promising research to meet their local needs and develop innovative resources and systems that we implement through multiple channels, from the highly local and personal to the regional and national. We provide long-term technical assistance, collaborate with partners at all levels of the education system, and advise community colleges and states.

We have significant experience and expertise in the following:

- Developing and implementing standards and building the capacity of schools, districts, and systems
- Supporting education leadership, instructional coaching, and teaching
- Designing and developing instructional materials, assessments, curricula, and programs for bridging critical transitions
- Convening networks focused on policy, research, and practice

The Center was founded in 1991 at The University of Texas at Austin. Our staff members have expertise in leadership, literacy, research, program evaluation, mathematics and science education, policy and systemic reform, and services to high-need populations. We have worked with states and education systems throughout Texas and across the country. For more information about our programs and resources, see our homepage at **www.utdanacenter.org**.

About the New Mathways Project

The NMP is a systemic approach to improving student success and completion through implementation of processes, strategies, and structures based on four fundamental principles:

1. Multiple pathways with relevant and challenging mathematics content aligned to specific fields of study
2. Acceleration that allows students to complete a college-level math course more quickly than in the traditional developmental math sequence
3. Intentional use of strategies to help students develop skills as learners
4. Curriculum design and pedagogy based on proven practice

The Dana Center has developed curricular materials for three accelerated pathways—Statistical Reasoning, Quantitative Reasoning, and Reasoning with Functions I and II (a two course preparation for Calculus). The pathways are designed for students who have completed arithmetic or who are placed at a beginning algebra level. All three pathways have a common starting point—a developmental math course that helps students develop foundational skills and conceptual understanding in the context of college-level course material.

In the first term, we recommend that students also enroll in a learning frameworks course to help them acquire the strategies—and tenacity—necessary to succeed in college. These strategies include setting academic and career goals that will help them select the appropriate mathematics pathway.

In addition to the curricular materials, the Dana Center has developed tools and services to support project implementation. These tools and services include an implementation guide, data templates and planning tools for colleges, and training materials for faculty and staff.

Acknowledgments

The development of the New Mathways Project curricular materials began with the formation of the NMP **Curricular Design Team**, who set the design standards for how the curricular materials for individual NMP courses would be designed. The team members are:

Richelle (Rikki) Blair, Lakeland Community College (Ohio)
Rob Farinelli, College of Southern Maryland (Maryland)
Amy Getz, Charles A. Dana Center (Texas)
Roxy Peck, California Polytechnic State University (California)
Sharon Sledge, San Jacinto College (Texas)
Paula Wilhite, Northeast Texas Community College (Texas)
Linda Zientek, Sam Houston State University (Texas)

The Dana Center then convened faculty from each of the NMP codevelopment partner institutions to provide input on key usability features of the instructor supports in curricular materials and pertinent professional development needs. Special emphasis was placed on faculty who need the most support, such as new faculty and adjunct faculty. The **Usability Advisory Group** members are:

Ioana Agut, Brazosport College (Texas)
Eddie Bishop, Northwest Vista College (Texas)
Alma Brannan, Midland College (Texas)
Ivette Chuca, El Paso Community College (Texas)
Tom Connolly, Charles A. Dana Center (Texas)
Alison Garza, Temple College (Texas)
Colleen Hosking, Austin Community College (Texas)
Juan Ibarra, South Texas College (Texas)
Keturah Johnson, Lone Star College (Texas)
Julie Lewis, Kilgore College (Texas)
Joey Offer, Austin Community College (Texas)
Connie Richardson, Charles A. Dana Center (Texas)
Paula Talley, Temple College (Texas)
Paige Wood, Kilgore College (Texas)

Some of the content for this course is derived from the Statway™ course, which was developed under a November 30, 2010, agreement by a team of faculty authors and reviewers contracted and managed by the Charles A. Dana Center at The University of Texas at Austin under sponsorship of the Carnegie Foundation for the Advancement of Teaching. Statway™ is copyright © 2011 by the Carnegie Foundation for the Advancement of Teaching and the Charles A. Dana Center at The University of Texas at Austin. Statway™ and Quantway™ are trademarks of the Carnegie Foundation for the Advancement of Teaching.

Funding and support for the New Mathways Project was provided by the Kresge Foundation, Carnegie Corporation of New York, Greater Texas Foundation, Houston Endowment, Texas legislative appropriations request, and TG.

Any opinions, findings, conclusions, or recommendations expressed in this material are those of the author(s) and do not necessarily reflect the views of these funders or The University of Texas at Austin. This publication was also supported through a collaboration between the Charles A. Dana Center, Texas Association of Community Colleges, and Pearson Education, Inc.

Acknowledgments for Version 1.0

Development of Version 1.0 (2015) of the *Quantitative Reasoning* course was made possible by a grant from the Greater Texas Foundation and the Houston Endowment.

Unless otherwise noted, all staff listed are with the Charles A. Dana Center at The University of Texas at Austin.

Project Lead, Authors, and Reviewers

Richelle (Rikki) Blair, lead author, professor emerita, Lakeland Community College (Ohio)
Scott Guth, professor, Mt. San Antonio College (California)
Rob Kimball, professor emeritus, Wake Tech Community College (North Carolina)
Andrea Levy, mathematics instructor, Seattle Central Community College (Washington)
Maura Mast, associate vice provost for undergraduate studies, University of Massachusetts Boston (Boston)
Aaron Montgomery, Professor, Central Washington University (Washington)
Jeff Morford, Henry Ford Community College (Michigan)
Victor Piercey, Assistant Professor, Ferris State University (Michigan)
Connie J. Richardson, project lead and lead author, *Quantitative Reasoning*
Jack Rotman, Lansing Community College (Michigan)
Jan Roy, mathematics department chair, Montcalm Community College (Michigan)
Erin Sagaskie, mathematics instructor, Carbondale Community High School (Illinois)

Project Staff

Adam Castillo, graduate research assistant
Heather Cook, program coordinator
Ophella C. Dano, lead editor
Amy Getz, strategic implementation lead, higher education
Monette McIver, manager, higher education services
Brandi M. Mendez, administrative assistant
Erica Moreno, program coordinator
Dana Peterson, independent consultant
Phil Swann, senior designer
Sarah Wenzel, administrative associate

Pearson Education, Inc. Staff

Vice President, Editorial Jason Jordan
Strategic Account Manager Tanja Eise
Editor in Chief Anne Kelly
Senior Acquisitions Editor Marnie Greenhut
Digital Instructional Designer Tacha Gennarino
Manager, Instructional Design Sara Finnigan
Senior Project Manager Dana Toney
Director of Course Production, MyMathLab Ruth Berry
MathXL Content Developer Bob Carroll
Project Manager Kathleen A. Manley
Project Management Team Lead Christina Lepre
Product Marketing Manager Alicia Frankel
Senior Author Support/Technology Specialist Joe Vetere
Rights and Permissions Project Manager Martha Shethar
Procurement Specialist Carol Melville
Associate Director of Design Andrea Nix
Program Design Lead Beth Paquin
Composition Dana Bettez

Contents

Lesson	Lesson Title and Description	
	Curriculum Overview	p. xiii
Lessons 1-7: Complex Numerical Summaries; Graphical Displays		
1.A	Data for Life *Collect data that will be referred to throughout the semester; supplemental spreadsheet provided*	p. 1
1.B	Our Learning Community *Student success focus* *Establish a sense of shared responsibility; provide key information about course content and policies*	p. 5
1.C	Instant Runoff *Voting schemes*	p. 7
1.D	Borda Count *Voting schemes*	p. 11
2.A	Graphical Displays *Analysis and communication; dotplots, histograms, boxplots; mean; median*	p.13
2.B	Forming Effective Study Groups *Student success focus* *Taking responsibility for own learning and supporting learning of others; setting norms*	p. 15
2.C	Mini-Project: Graphical Displays *Write formal, contextual analysis on compared data; research-related data; sample rubric provided*	p. 17
3.A	Who Is in the Population? *Populations; sampling*	p. 19
3.B	How Much Water Do I Drink? *Analyzing class data; Central Limit Theorem*	p. 21
3.C	How Much Water Does Our Class Drink? *Sample standard deviation*	p. 23
4.A	What Are the Risks? *Theoretical probability of two or more independent events*	p. 27
4.B	Calculating Risk *Conditional probability of two or more dependent events*	p. 31

Lesson	Lesson Title and Description	
5.A	Cost of Living Comparisons *Conversion to create equivalent units; supplemental spreadsheet*	p. 35
5.B	Index Numbers *Using indices such as Consumer Price Index; supplemental spreadsheet*	p. 39
5.C	Polls, Polls, Polls! *Weighted averages*	p. 43
5.D	Average Income *Weighted averages and expected value; supplemental spreadsheet*	p. 47
6.A	How Can We Smooth the Data? *Simple and weighted moving averages; supplemental spreadsheet*	p. 49
6.B	Mini-Project: Income Disparities *Written analysis of graphical display of weighted moving average*	p. 51
7.A	U.S. Budget Priorities *Part-part vs. part-whole ratios*	p. 53
7.B	Understanding U.S. Budget Priorities *Decimals, percentages, and part-whole ratios*	p. 57
7.C	Changes to U.S. Budget Priorities *Absolute and relative change*	p. 61
7.D	Percent of Total U.S. Budget *Dotplots used to introduce symmetry and skewness*	p. 63
7.E	What's My Credit Score? *Application of ratios*	p. 65
7.F	U.S. Incarceration Rates *Applications of ratios; comparison*	p. 69
Lessons 8-12: Mathematical Modeling		
8.A	More Water, Please! *Introduction to mathematical modeling*	p. 71
8.B	What's My Car Worth? *Distinguishing proportionality and linearity*	p. 75
8.C	How Money Makes Money *Non-linear models*	p. 79
8.D	Have My Choices Affected My Learning? *Regression using student data*	p. 83

Lesson	Lesson Title and Description	
16.E	Big Data (GIS) *Analysis problems associated with large, volatile data*	p. 229
16.F	Big Brother – They're Watching! *Conclusions from heat maps*	p. 237
17.A	Decisions, Decisions *Decision making based on multiple pieces of quantitative information*	p. 241
17.B	The Write Approach to Data *Improving written analyses of graphical displays*	p. 245
17.C	Numbers Never Lie *Misleading and erroneous graphical displays*	p. 249
17.D	Can You Feel the Heat? *Using data to understand complex issues*	p. 255
18.A	Mini-Project: Tornado Climatology *Choosing appropriate ways to represent data*	p. 259
18.B	The Making of a Model *Various ways to present mathematical models*	p. 263
18.C	What a Wonderful World! *Using multiple representations to choose a model*	p. 267
18.D	Mathematical Models *Limitations of models*	p. 271

Student Resources

Curriculum Overview

Contents

About *Quantitative Reasoning*

Quantitative Reasoning (*QR*) is designed for students who have completed *Foundations of Mathematical Reasoning* (*Foundations* or *FMR*) and the co-requisite *Frameworks for Mathematics and Collegiate Learning* (*Frameworks* or *FMCL*). If you did not take one or either of these courses, see your instructor about strategies for acclimating to the structure and climate of this course.

Quantitative Reasoning will help you develop quantitative literacy skills that will be meaningful for your professional, civic, and personal lives. Such reasoning is a habit of mind, seeking pattern and order when faced with unfamiliar contexts. In this course, an emphasis is placed on the need for data to make good decisions and to have an understanding of the dangers inherent in basing decisions on anecdotal evidence rather than on data.

Structure of the curriculum

The *QR* curriculum is designed in 25-minute learning episodes, which can be taught in one, two, three, or more lesson groups to conform to the class length. These short bursts of active learning, combined with whole class discussion and summary, produce increased memory retention.[1]

---

[1] Sources: Buzan, T. (1989). *Master your memory* (Typersetters); Buzan, T. (1989). *Use your head* (BBC Books); Sousa, D. (2011). *How the brain learns, 4th ed.* (Corwin); Gazzaniga, M., Ivry, R. B., & Mangun, G. R. (2002). *Cognitive neuroscience: The biology of the mind, 2nd ed.* (W.W. Norton); Stephane, M., Ince, N., Kuskowski, M., Leuthold, A., Tewfik, A., Nelson, K., McClannahan, K., Fletcher, C., & Tadipatri, V. (2010). Neural oscillations associated with the primary and recency effects of verbal working memory. *Neuroscience Letters*, 473, 172–177); Thomas, E. (1972). The variation of memory with time for information appearing during a lecture. *Studies in Adult Education,* 57–62.

QR continues the philosophy and structures of the *Foundations* course. The lesson structure includes a Preview Assignment (to be completed before class); the Lesson (Student Pages); and a Practice Assignment (to be completed after class). Some lessons do not have a Preview or Practice, but all Lessons include Student Pages.

Constructive perseverance level

You may or may not have experience with struggling in mathematics. Struggle is important, because struggling indicates learning. If struggle is not taking place, you are not being challenged and are not gaining new knowledge and skills. However, struggle that is unproductive often turns into frustration. The *Quantitative Reasoning* course is designed to promote constructive perseverance—that is, you are supported in persisting through struggle.

The levels of constructive perseverance, outlined below, should be viewed as a broad continuum rather than as distinct, well-defined categories. Some content requires greater structure and more direct instruction. The levels are not designated in your materials, but the descriptions may help you understand the structure of the lessons. The levels of constructive perseverance are as follows:

- **Level 1:** The problem is broken into sub-questions that help develop strategies. You and your fellow group members reflect on and discuss questions briefly and then come back together to discuss with the full class. This process moves back and forth between individual or small group discussion and class discussion in short intervals.

  Role of the instructor: To help the class develop the culture of discussion, establish norms of listening, and model the language used to discuss quantitative concepts.

- **Level 2:** The problem is broken into sub-questions that give you some direction but do not explicitly define or limit strategies and approaches. You work in groups on multiple steps for longer periods, and the instructor facilitates with individual groups, as needed. The instructor brings the class together at strategic points to make important connections explicit or when breakdowns of understanding have occurred or are likely to occur.

  Role of the instructor: To support you in working more independently and evaluating your own work so you feel confident about moving through multiple questions without constant reinforcement from the instructor.

- **Level 3:** The problem is not broken into steps or is broken into very few steps. You are expected to identify strategies for yourself. Groups work independently on the problem with facilitation by the instructor, as necessary. Groups report on results, and class discussion focuses on reflecting on the problem as a whole.

  Role of the instructor: To support students in persisting with challenging problems, including trying multiple strategies before asking for help.

The role of the Preview and Practice Assignments

One of the most important aspects of the *Quantitative Reasoning* curriculum is the role and design of the homework assignments (Preview Assignments before the lesson and Practice Assignments after the lesson). These assignments differ from traditional homework in several ways.

- The Preview Assignments occasionally contain information or activities (internet search or review, view a video, read an article, etc.) that are used in the next lesson.

- The Preview Assignments are designed to prepare you for the next lesson and to review mathematical concepts that will be needed. You are given a set of skills for the next lesson and asked to rate yourself on your readiness to use those skills. Be honest when rating yourself—if you are unprepared for class, you will be unable to participate fully. You are instructed to seek help before the next class meeting if you are unable to successfully complete these problems.

Monitoring Your Readiness

To effectively plan and use your time wisely, it helps to think about what you know and do not know. For each of the following, rate how confident you are that you can successfully do that skill. Use the following descriptions to rate yourself:

5–I am extremely confident I can do this task.
4–I am somewhat confident I can do this task.
3–I am not sure how confident I am.
2–I am not very confident I can do this task.
1–I am definitely not confident I can do this task.

Use the ratings to get ready for the upcoming in-class activity. Remember, your instructor is going to assume that you are confident with the material and will not take class time to answer questions about it.

How confident are you that you can estimate the double of a number?

Choose.....

5—I am extremely confident I can do this task.
4—I am somewhat confident I can do this task.
3—I am not sure how confident I am.
2—I am not very confident I can do this task.
1—I am definitely not confident I can do this task.

t between numbers and number-word combinations?

How confident are you that you can read a timeline?

Choose.....

- The Practice Assignments provide you with opportunities to develop, practice, and extend skills from the current lesson. These assignments may include similar problems in a new context or an extension of the learning within the same context.

- One goal of the course is for you to engage increasingly in productive struggle. Therefore, the assignments are based on the same principle of constructive perseverance as the rest of the curriculum. Ideally, each assignment should offer entry-level questions that you should be able to complete successfully, followed by more challenging questions. You should make a valid attempt on each question, even if you are not sure about the work.

Resource materials for students

The Student Resources are designed to be a starting point for course reference materials. Consider adding to them as you view videos in MyMathLab or research information on the internet.

Language and literacy skills

Quantitative literacy has unique language demands that are distinct from other subjects, including other math courses.

The purpose of writing in the *Quantitative Reasoning* course is to:

- Make sense of quantitative information and processes, especially in relationship to a context.
- Build skills in communicating about quantitative information.
- Provide a form of assessment by which you demonstrate your understanding of the course material in writing.

Curriculum design standards

The New Mathways Project (NMP) is made up of individual courses that form *pathways* to take you to and through college-level mathematics. The concept of the pathway as a yearlong experience is critical to the NMP because these courses are designed to articulate in a way that provides you with the experience of learning mathematics and/or statistics through coherent, consistent practices and structures.

The design standards outlined in this section set the guidelines for how the curricular materials for individual NMP courses are designed to support that coherent experience for you.

Note: The numbering in the description of the design standards does not indicate level of importance.

Standard I: Structure and Organization of Curricular Materials

The NMP is organized around big mathematical and statistical ideas and concepts as opposed to skills and topics.

Standard II: Active Learning

The NMP is designed to actively involve you in doing mathematics and statistics, analyzing data, constructing hypotheses, solving problems, reflecting on their work, and learning and making connections.

Class activities provide regular opportunities for you to actively engage in discussions and tasks using a variety of different instructional strategies (e.g., small groups, class discussions, interactive lectures).

Standard III: Constructive Perseverance

The NMP supports you in developing the tenacity, persistence, and perseverance necessary for learning mathematics.

Standard IV: Problem Solving

The NMP supports you in developing problem-solving skills, and students apply previously learned skills to solve nonroutine and unfamiliar problems.

Standard V: Context and Interdisciplinary Connections

The NMP presents mathematics and statistics in context and connects mathematics and statistics to various disciplines.

Standard VI: Use of Terminology

The NMP uses discipline-specific terminology, language constructs, and symbols to intentionally build mathematical and statistical understanding and to ensure that terminology is not an obstacle to understanding.

Standard VII: Reading and Writing

The NMP develops your ability to communicate about and with mathematics and statistics in contextual situations appropriate to the pathway.

Standard VIII: Technology

The NMP uses technology to facilitate active learning by enabling students to directly engage with and use mathematical concepts. Technology should support the learning objectives of the lesson. In some cases, the use of technology may be a learning objective in itself, as in learning to use a statistical package in a statistics course.

Note: A more detailed description of the design standards is available on the Dana Center website at http://www.utdanacenter.org/higher-education/new-mathways-project/new-mathways-project-curricular-materials/foundations-of-mathematical-reasoning-course/

Readiness competencies

The skills listed below will help you to succeed in *Quantitative Reasoning*.

- Demonstrate procedural fluency with real number arithmetic operations and use those operations to represent real-world scenarios and to solve stated problems. Demonstrate number sense, including dimensional analysis and conversions between fractions, decimals, and percentages. Determine when approximations are appropriate and when exact calculations are necessary.

- Solve linear equations, graph and interpret linear models, and read and apply formulas.
- Demonstrate a basic understanding of displays of univariate data such as bar graphs, histograms, dotplots, and circle graphs, including appropriate labeling.
- Take charge of your own learning through good classroom habits, time management, and persistence. Participate in the classroom community through written and oral communication.

NMP Pathway learning goals with applications for the *QR* course

The following five learning goals apply to all NMP mathematics courses, with the complexity of problem-solving skills and use of strategies increasing as you advance through your pathway.

For each NMP course, we define the ways that the learning goals are applied and the expectations for mastery. The bullets below each of the five learning goals specify the ways in which each learning goal is applied in the *Quantitative Reasoning* course.

Each NMP course is designed so that you meet the goals across the courses in your pathway. Within a course, the learning goals are addressed across the course's content-based learning outcomes.

Communication Goal: You will be able to interpret and communicate quantitative information and mathematical and statistical concepts using language appropriate to the context and intended audience.

In the *Quantitative Reasoning* course, you will...

- Use appropriate mathematical and statistical language in oral, written, and graphical forms.
- Read and interpret authentic texts such as advertisements, consumer information, government forms, and newspaper articles containing quantitative information, including graphical displays of quantitative information. These texts may be as long as a standard magazine article and will include comparisons, analysis, and synthesis of multiple forms or sources of quantitative information.
- Write 1 to 2 pages using quantitative information to synthesize information from multiple sources or to make or critique an argument.

Problem Solving Goal: You will be able to make sense of problems, develop strategies to find solutions, and persevere in solving them.

In the *Quantitative Reasoning* course, you will...

- Develop an answer to an open-ended question requiring analysis and synthesis of multiple calculations, data summaries, and/or models. You will be expected to develop your own process with support from peers and the instructor. This type of question would be expected to extend over time (beyond one or two class meetings) with work occurring both in class and outside of class with specific checkpoints to monitor progress.

Reasoning Goal: You will be able to reason, model, and draw conclusions or make decisions with mathematical, statistical, and quantitative information.

In the *Quantitative Reasoning* course, you will...

- Draw conclusions or make decisions in quantitatively based situations that are dependent upon multiple factors. You will analyze how different situations would affect the decisions.
- Present written or verbal justifications of decisions that include appropriate discussion of the mathematics involved.
- Recognize when additional information is needed.

Evaluation Goal: You will be able to critique and evaluate quantitative arguments that utilize mathematical, statistical, and quantitative information.

In the *Quantitative Reasoning* course, you will...

- Evaluate the validity and possible biases in arguments presented in authentic contexts based on multiple sources of quantitative information (e.g., advertising, internet postings, consumer information, political arguments).

Technology Goal: You will be able to use appropriate technology in a given context.

In the *Quantitative Reasoning* course, you will...

- Use a spreadsheet to organize quantitative information and make repeated calculations using simple formulas.
- Use the internet to find quantitative information on a given subject and evaluate the validity and possible bias of information based on the source.
- Use internet-based tools appropriate for a given context (e.g., an online tool to calculate credit card interest).

Content learning outcomes for *Quantitative Reasoning*

The topics for the *Quantitative Reasoning* course are:

- Number, Ratio, and Proportional Reasoning
- Modeling
- Probability
- Statistics

Number, Ratio, and Proportional Reasoning

Outcome: You will draw conclusions and/or make decisions based on analysis and critique of quantitative information using proportional reasoning. You will also effectively justify conclusions and communicate about their conclusions in ways appropriate to the audience.

You will be able to:

N.1 Solve real-life problems requiring interpretation and comparison of complex numeric summaries which extend beyond simple measures of center.

For example: Interpret and/or compare weighted averages, indices, coding, ranking; evaluate claims based on complex numeric summaries.

N.2 Solve real-life problems requiring interpretation and comparison of various representations of ratios, (i.e. fractions, decimals, rate, and percentages).

For example: Interpret non-standard ratios used in media and risk reporting; identify and contrast in written statements part-to-part versus part-to-whole ratios taken from meaningful context; understand and communicate percentages as rates per 100; identify uses and misuses of percentages related to a proper understanding of the base; analyze growth and decay using absolute and relative change and comparisons using absolute and relative difference.

N.3 Distinguish between proportional and non-proportional situations, and, when appropriate, apply proportional reasoning.

For example: Solve for an unknown quantity in proportional situations; determine the constant of proportionality in proportional situations, leading to a symbolic model for the situation (i.e., an equation based upon a rate of change, ($y = kx$); solve real-life problems requiring conversion of units using dimensional analysis; apply scale factors to perform indirect measurements (e.g., maps, blueprints, concentrations, dosages, densities); recognize when proportional techniques do not apply.

Modeling

Outcome: You will draw conclusions and/or make decisions by analyzing and/or critiquing mathematical models, including situations for which you must recognize underlying assumptions and/or make reasonable assumptions for the model.

You will be able to:

M.1 Analyze and critique mathematical models and be able to describe their limitations.

For example: Distinguish between correlation and causation; determine whether interpolation and/or extrapolation are appropriate.

M.2 Use models, including models created with spreadsheets or other tools, to estimate solutions to contextual questions, identify patterns and identify how changing parameters affect the results.

For example: Functional models to estimate future population; spreadsheets to model financial applications (e.g., credit card debt, installment savings, amortization schedules, mortgage and other loan scenarios).

M.3 Choose and create models for bivariate data sets, and use the models to answer questions and make decisions.

For example: Determine whether data can best be modeled by a linear, exponential, logistic, or periodic function; create models by hand or with technology; use models appropriately; demonstrate understanding of the limitations of chosen models.

Probability

Outcome: You will apply probabilistic reasoning to draw conclusions, to make decisions, and to evaluate outcomes of decisions.

You will be able to:

P.1 Evaluate claims based on empirical, theoretical, and subjective probabilities.

For example: Analyze outcomes and make decisions related to risk, pay-off, expected value, and false negatives/positives in various probabilities contexts.

P.2 Use data displays and models to determine probabilities (including conditional probabilities) and use these probabilities to make informed decisions.

For example: Two-way tables, tree diagrams, Venn diagrams, and area models.

Statistics

Outcome: You will draw conclusions or make decisions and communicate their rationale based on understanding, analysis, and critique of self-created or reported statistical information and statistical summaries.

You will be able to:

S.1 Use statistical information from studies, surveys, and polls (including when reported in condensed form or summary statistics) to make informed decisions.

For example: Identify limitations, strengths, or lack of information in studies, including data collection methods (e.g., sampling, experimental, observational) and possible sources of bias; identify errors or misuses of statistics to justify particular conclusions; interpret and compare the results of polls using margin of error.

S.2 Create and use visual displays of data.

For example: Create (with and without technology) visual representations of real-world data, such as charts, tables and graphs; interpret and analyze visual representations of data; describe strengths, limitations, and fallacies of various graphical displays.

S.3 Summarize, represent, and interpret data sets on a single count or measurement variable.

For example: Use plots and statistics appropriate to the shape of the data distribution to represent a single data set; compare center, shape, and spread of two or more data sets; interpret the differences in context.

S.4 Use properties of distributions to analyze data and answer questions.

For example: Recognize when data are normally distributed and use the mean and standard deviation of the data to fit to a normal distribution.

Lesson 1, Part A Data for Life

This class is a community of learners. Having shared goals strengthens communities. Our shared goal for this course is to maximize the learning experience for everyone. To do that, we will begin by working together to collect data that will be organized and used in later lessons. We will also reflect on activities that influence our learning.

Credit: Monkey Business/Fotolia

1) Complete question 1 on the Data Recording Sheet (Mandatory Categories).

Objectives for the lesson

You will understand that:

- ☐ Data can be collected from many real-life sources and used for future mathematical study.
- ☐ Analysis of data can provide information to help make decisions about positive changes in your study habits and lives.
- ☐ Working together builds a positive learning community.

You will be able to:

- ☐ Collect data from your daily life.
- ☐ Work positively in a group to make a decision.

2) Now look at question 2 on the Data Recording Sheet (Optional Categories). We need to narrow down the list to three categories. Rate the categories by putting them in order from most likely to affect your ability to learn (#1) to least likely to affect your learning (#9). Do not fill in the last column yet.

> **Consensus** is an agreement. It is not a vote. When we vote, some people win and others lose. With consensus, the group comes to an agreement that everyone supports.

3) Discuss your priorities with your group and come to consensus on your group's top three choices. A consensus means that you all agree. In a community of learners, everyone wins.

4) How did your group make your choices and reach consensus?

5) As we begin the course and create a positive learning environment, let's reflect on activities that can positively affect your ability to learn. List three activities that would have a positive effect on your learning in this class.

6) Now reflect on activities that could negatively affect your ability to learn. List three activities that would have a negative effect on your learning in this class.

7) List one action that you could take to minimize the negative effects on your learning.

8) Collect your data on daily water consumption for the next 10 days on the Daily Water Consumption Data Recording Table below. You will need this information for a later lesson.

Daily Water Consumption Data Recording Table

Day	1	2	3	4	5	6	7	8	9	10
Number of cups of water consumed per day										

Data Recording Sheet

1) Complete the information for the following Mandatory Categories table below.

Mandatory Categories	Record your personal information.
Gender (male or female)	
Height, expressed as a decimal in feet and inches (For example: 5' 7" = about 5.58 feet)	
Number of hours per week working on math with others outside of class (tutoring, study group)	
Number of hours per week working on math alone outside of class	
Number of hours per week spent in math class	
Number of cups of water consumed per day	
Number of hours per week commuting to work and college	

2) Consider the Optional Categories table below. Number the categories from most likely to affect your learning (#1) to least likely to affect your learning (#9). Discuss your ordering with your group members. Then the class will come to consensus on the top three categories for the class. These categories will be used as data sets in future lessons. When the top three categories are identified, record your own data for those three categories. Turn in this sheet to your instructor at the end of class.

Optional Categories	Number your choices.	Record your personal information for the top three categories chosen by the class.
Number of alcoholic beverages consumed per week		
Number of caffeinated beverages consumed per week		
Number of daily servings of fruits and vegetables		
Number of daily servings of protein		
Number of hours exercising (e.g., walking, yoga, biking, sports) per week		
Number of hours using technology for fun per week (e.g., TV, email, video games, texting, Facebook)		
Number of hours working per week		
Number of hours sleeping per week		
Number of hours preparing meals and eating per week		

Lesson 1, Part B Our Learning Community

Building a strong learning community can help everyone achieve at a higher level in this class.

Credit: shock/Fotolia

1) What are some things each of us can do to contribute to our learning community?

Objectives for the lesson

You will understand:

- ☐ Your importance in the learning community.
- ☐ The course policies and procedures you need to know and follow.

You will be able to:

- ☐ Seek and give help to one another inside and outside of class.

2) Use your syllabus and conversations with classmates to complete the information on these pages. Use them as a resource throughout the semester, adding to them when you find new avenues of support.

Instructor's Name

Instructor's Office Location

Instructor's Office Hours

Office Phone Number

Email Address

Tutoring Lab Location and Phone Number

Computer Lab Location

Other On-campus Resources

Classmate(s) Contact Information

Lesson 1, Part C Instant Runoff

Elections are held every day to choose persons for public office. Election methods can also be used to select between multiple scenarios or issues, such as the menu for an office party.

1) In an election involving two people, when looking at the votes cast, what criteria would you use to determine who should win? What about an election involving three people?

Credit: flysnow/Fotolia

Objectives for the lesson

You will understand:

- ☐ That earning the most votes may not be sufficient to win an election.
- ☐ That there are multiple considerations and methods for ranking candidates in an election.
- ☐ That multiple ranking methods can be employed to make decisions about other issues.
- ☐ The difference between the terms plurality and majority in an election.

You will be able to:

- ☐ Create a first-degree equation involving percentages and solve for the variable.
- ☐ Employ the "Instant Runoff" method to determine the winner of an election.
- ☐ Apply and justify selection strategies to election results and decisions about other issues.

A candidate (or proposal) who receives more than half of the votes in an election is said to have won a **majority** or an **absolute majority.** In an election with more than two candidates, the candidate who receives the largest number of votes (but not necessarily more than half) is said to have received a **plurality** of the votes. A plurality is sometimes called a **relative majority.**

There are several ways to determine a winner in situations where no candidate received a majority of the votes. This activity and the next provide some examples.

2) In 2012, one of Texas' U.S. Senate seats was up for grabs. Nine candidates were on the ballot for the Republican Party primary. The results of the primary are shown below.[1] Who do you think should have won? Why?

Candidate	Glenn Addison	Joe Agris	Curt Cleaver	Ted Cruz	David Dewhurst	Ben Gambini	Craig James	Tom Leppert	Lela Pittenger
% of Vote	2%	0%	0%	34%	45%	1%	4%	13%	1%

3) Texas election rules state that elections are majority elections. Who is the winner of this election?

4) The number of votes received by Cruz and Dewhurst are listed below.

Candidate	Original Primary Votes (%)
Ted Cruz	477,428 (34%)
David Dewhurst	621,850 (45%)

Part A: How many total votes were cast in the election?

Part B: How many votes would be needed to win in a majority election?

5) Since no candidate received a majority in the original primary election, the election rules required a runoff election. The results of the runoff election are shown below. The runoff is also a majority election. Who won the runoff election? How?

[1] Source: http://www.thepoliticalguide.com/Elections/2012/Senate/Texas/1/

Candidate	**Runoff Votes (%)**
Ted Cruz	628,336 (57%)
David Dewhurst	477,888 (43%)

Another election method called the "Instant Runoff" method, or elimination method, is to ask voters to rank the candidates in order of preference. Consider the following scenario:

Three candidates (Alex, Blake, and Charlie) applied for a position at a company. The interviewing committee ranked their choices in order of preference on a ballot. The completed ballot from one committee member is shown.

	List the candidates in order of your preference
1st choice	Alex
2nd choice	Blake
3rd choice	Charlie

6) Once all of the ballots are collected, the results are compiled in a preference schedule. Notice that three people ranked Alex in 1st place and Blake in 2nd place, and so on.

	3 voters	1 voter	2 voters	1 voter	2 voters
1st choice	Alex	Alex	Blake	Blake	Charlie
2nd choice	Blake	Charlie	Alex	Charlie	Blake
3rd choice	Charlie	Blake	Charlie	Alex	Alex

Part A: How many voters were on the interviewing committee?

Part B: A majority is still needed to win in his scenario. How many votes are needed for a majority?

Part C: How many 1st place votes did each candidate receive? Do we have a winner?

Part D: Which candidate received the fewest 1st place votes? To employ the Instant Runoff method, mark out or eliminate that person's name in each column of the table. Any name below that one in the table will now move up.

Part E: How many 1st place votes do we now award to the two remaining candidates? Who wins the election?

7) The method shown in question 6 is known as the Instant Runoff method. What are some pros and cons of this method?

8) Can you think of another election method for choosing the winning candidate in an election? What are pros and cons of your method? Can any process guarantee that the choice made (or winner) in an election is correct and fair to all candidates?

Lesson 1, Part D Borda Count

Recall that in Lesson 1, Part C, an interviewing committee was trying to select the appropriate applicant to fill a job position. In this lesson, we will explore another election method called the Borda Count method.

Credit: flysnow/Fotolia

1) Suppose the committee's preference schedule looks like the one shown below.

Part A: How many people are on the committee? How many votes are needed for a majority?

Part B: Does anyone have a majority of the votes?

	1 voter	3 voters	3 voters	1 voter
1st choice	Alex	Alex	Blake	Blake
2nd choice	Blake	Charlie	Alex	Charlie
3rd choice	Charlie	Blake	Charlie	Alex
Column 1	**Column 2**	**Column 3**	**Column 4**	**Column 5**

Objectives for the lesson

You will understand:

- ☐ Earning the most votes may not be sufficient to win an election.
- ☐ There are multiple considerations and methods for ranking candidates in an election.
- ☐ Multiple ranking methods can be employed to make decisions about other issues.

You will be able to:

- ☐ Employ the Borda Count method to determine the winner of an election.
- ☐ Apply and justify selection strategies to election results.

2) In the Borda Count method, voters' choices are given **weights**. In an election with 3 choices, a voter's 1st choice is worth 3 points, the 2nd choice is worth 2 points, and the 3rd choice is worth 1 point.

Part A: Write these point values in the appropriate cells of the first column.

Part B: Column 2 represents one voter. How many points did that voter give Alex? Blake? Charlie? Write the appropriate points next to each applicant's name in Column 2.

Part C: In Column 3, how many total points did Alex receive? (Notice that three voters put Alex in 1st place, each giving him 3 points.) How many points did Charlie and Blake receive? Fill in Column 3, then continue to Columns 4 and 5.

Part D: How many total points did each candidate receive? Who should be offered the position?

Part E: If that person already took another job and turns down the offer, should the committee offer the job to someone else? If so, who and why?

Part F: How many points does each voter award to candidates? How many total points are then available to candidates? How can you use this information to cross-check your work?

3) What advantages or disadvantages does the "point" method of the Borda Count election method provide compared to the elimination method (Instant Runoff) from Lesson 1, Part C? Explain your answer.

Lesson 2, Part A Graphical Displays

In Preview Assignment 2.A, you analyzed three graphical displays that represented the same data on treadmill ratings.

Credit: shock/Fotolia

1) What are characteristics of the dotplot, histogram, and boxplot? Why do we need different types of graphical displays?

Objectives for the lesson

You will understand that:

- ☐ The same data can be displayed in different kinds of graphs.
- ☐ Different graphical displays have different purposes.
- ☐ Displays can be used to represent one data set or to compare two sets.

You will be able to:

- ☐ Analyze a variety of graphical displays and interpret them in context.
- ☐ Compute the mean of a set of data.
- ☐ Construct a dotplot or histogram from data.

2) The Consumer Reports article described in the Preview Assignment not only provided score ratings on treadmills, but also separated the treadmills into two cost categories: those costing $1,000 or less, and those costing more than $1,000.[1]

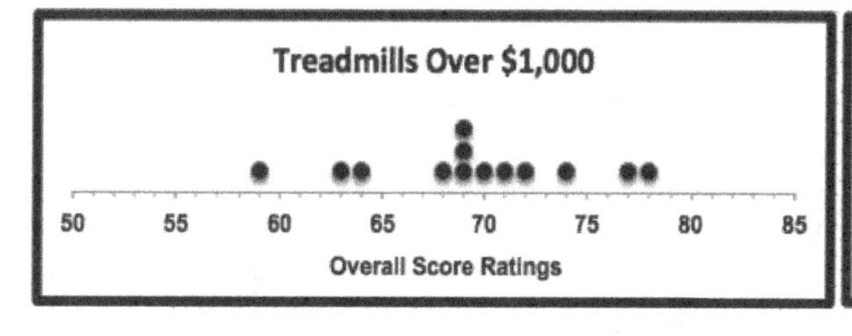

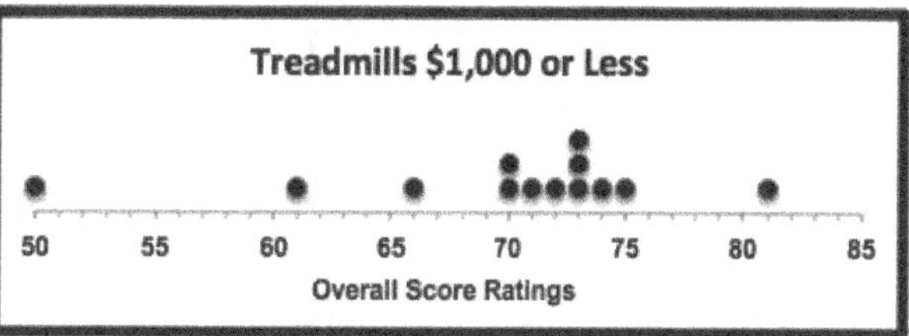

[1] Consumer Reports. (2011.) "Treadmills & ellipticals: 15 machines are tops in our latest tests." Retrieved November 1, 2013, from http://www.consumerreports.org/cro/magazine-archive/2011/february/home-garden/treadmills-ellipticals/overview/index.htm.

Part A: Compare and contrast the graphs of the two categories.

Part B: What conclusions do you draw from analyzing the two graphs?

Part C: Would a histogram or a boxplot be a useful method of displaying this data?

3) Suppose your friend is going to purchase a treadmill, and he has plenty of money but a limited amount of time for shopping. Would you recommend that he focus his attention on more expensive models or the less expensive models? Be specific as you explain your reasoning.

4) Review the data collected on the first day of class.

Part A: Construct a dotplot of the data of the height of the students in the class.

Part B: Compute the mean of the data set.

Part C: What conclusions can you draw from your calculations and the graph?

Lesson 2, Part B Forming Effective Study Groups

Our discussion about creating a positive learning environment and learning community continues from Lesson 1, Part B. One component of a productive learning community can be out-of-class study groups.

Credit: Syda Productions/Fotolia

1) What is a **study group** and what is its purpose?

Objectives for the lesson

You will understand that:

- ☐ Successful students learn best when they are actively involved in the learning process.
- ☐ Successful students working in small study groups tend to learn more and retain information longer than when the same content is presented in other instructional formats.
- ☐ Students have a shared responsibility in a learning community.

You will be able to:

- ☐ Describe how to form and conduct an effective study group.
- ☐ Identify key characteristics of effective study groups.
- ☐ Form a study group and become an active member of the group.

2) Consider the following statement:

Studying in groups helps students learn more effectively.

Part A: Do you agree with this statement? Why or why not?

Part B: Have you ever participated in a study group that met outside of a college class?

Part C: In the past, have you had a positive or negative experience with a study group?

Part D: What are some reasons for that experience?

3) Consider some additional questions related to study groups.

Part A: What are some reasons students would consider participating in a study group?

Part B: What are some guidelines to forming and conducting a successful study group?

Part C: What are some ways in which the group could determine its meeting time?

4) Identify 3 to 4 classmates and agree to form a study group.

Part A: Agree on a location/day/time of the first meeting, how many times a week you will meet, who will serve as an initial leader/facilitator, and what will be discussed at the first meeting.

Part B: Write down the name, phone number, and email address of each study group member.

Part C: Turn in the names of your study group members to your instructor.

Lesson 2, Part C Mini-Project: Graphical Displays

Our mini-project involves an analysis of data of median household incomes of Democrats and Republicans in the 2008 presidential election.

Credit: doganmesut/Fotolia

1) Why do people choose particular political parties?

2) Would higher income be more likely to be associated with the Democratic or Republican party?

Objectives for the lesson

You will understand that:

- ☐ Graphical displays can be used to compare two sets of data.

You will be able to:

- ☐ Research additional related data and look for trends in the data (optional).
- ☐ Write a contextual analysis of a graphical display in a formal paper at least two paragraphs long, including appropriate mathematical language and explanations.

The table on the next page lists the median household income (in thousands of dollars) for each state in the 2008 presidential election. The states are separated into two groups: those that were carried by Barack Obama and those carried by John McCain.

3) Consider the following excerpt of the table. What does this excerpt tell us about California's voters? What other information is given in the table?

States Won by Obama	Median Household Income (Thousands of Dollars)
California	57

4) Data for the 50 states and side-by-side boxplots of the data are shown. Your mini-project assignment is to write a paper of at least two paragraphs, comparing and contrasting the boxplots of the income of the states that voted for each of the candidates. This formal paper will be graded according to the rubric distributed by your instructor.

States Won by Obama (Democratic Party)	Median Household Income (Thousands of Dollars)
California	57
Colorado	61
Connecticut	65
Delaware	51
District of Columbia	56
Florida	45
Hawaii	62
Illinois	53
Indiana	47
Iowa	50
Maine	47
Maryland	64
Massachusetts	60
Michigan	50
Minnesota	55
Nevada	55
New Hampshire	66
New Jersey	65
New Mexico	42
New York	50
North Carolina	43
Ohio	47
Oregon	52
Pennsylvania	51
Rhode Island	53
Vermont	51
Virginia	62
Washington	57
Wisconsin	51

States Won by McCain (Republican Party)	Median Household Income (Thousands of Dollars)
Alabama	44
Alaska	64
Arizona	47
Arkansas	40
Georgia	46
Idaho	47
Kansas	48
Kentucky	41
Louisiana	40
Mississippi	36
Missouri	46
Montana	43
Nebraska	51
North Dakota	50
Oklahoma	46
South Carolina	42
South Dakota	52
Tennessee	40
Texas	46
Utah	63
West Virginia	38
Wyoming	53

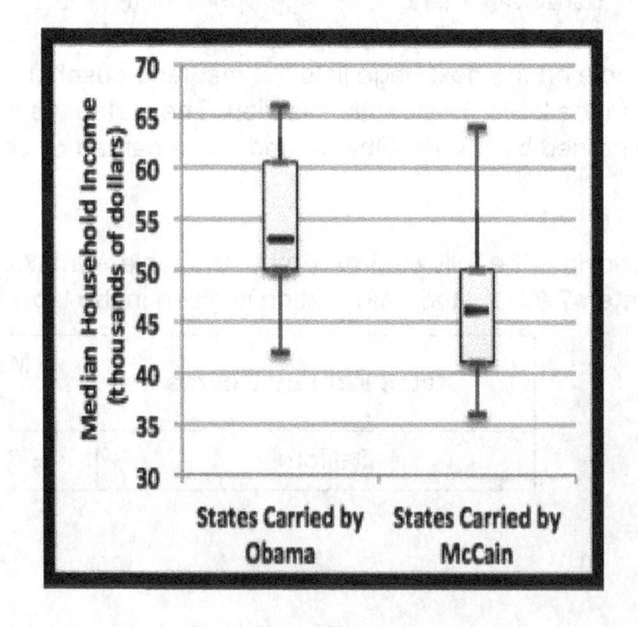

Lesson 3, Part A Who Is in the Population?

Lessons 3 and 4 focus on populations, samples, complex numerical summaries, and probabilities of research studies.

Credit: Stephen Finn/Fotolia

1) Jot down some "research studies" that you have heard about.

2) Recall the three polling groups that were presented in the Preview Assignment. When a company, such as Gallup, conducts a Presidential approval poll, why don't they contact everybody in the country?

Objectives for the lesson

You will understand:

- ☐ Why sampling is necessary in a statistical study.
- ☐ That limiting the characteristics of the sample also limits the conclusions one can draw about the population.

You will be able to:

- ☐ Explain the difference between a population and a sample.
- ☐ Use the characteristics of a study sample to describe the population.
- ☐ Analyze the conclusions of a study and explain the limitations on any inferences made about the population.

Researchers conducting studies usually need to draw a **sample** from a **population**. When a question about a group for a study is designed, the **population** is the entire group that they would like to say something about. The **sample** is a smaller group selected from the population that they want to study (a subset of the population). The data for the study is collected from the sample.

3) Let's conduct a research study about this class. What is the ratio of males to females in this class?

4) Do you think that this ratio is the same ratio in all classes across the entire campus? Why or why not?

5) In our study about the male-to-female ratio, describe the population and the sample.

Inference is the process of using data from a random sample to draw conclusions about a population. Although we cannot obtain the exact value of a measurement about a population from a sample, we can "get close." We use statistical techniques to determine whether we should depend on our results.

6) In our study about the male-to-female ratio, what is the inference we made about the ratio in our class related to the ratio of the institution?

7) Consider the following examples. Describe the population in each scenario from which the sample was drawn.

Part A: A random sample of 2,000 adults in Florida was asked if they support legalization of marijuana for medical use.

Part B: A random sample of 1,250 registered voters in Grand Rapids, Michigan was asked if they approve of a property tax increase to fund additional school maintenance.

Part C: The Nielson ratings company randomly selects 2,500 families from across the United States to participate in their television ratings program.

8) In each of the examples from question 7, describe the inference that would be made in this study.

Lesson 3, Part B How Much Water Do I Drink?

In Preview Assignment 3.B, you read an article about the importance of water for our health. In the previous lesson, you learned that samples could be used to infer information about a population.

Credit: BillionPhotos.com/Fotolia

1) Do you feel your classmates are a representative sample regarding water consumption for all students at your college?

Objectives for the lesson

You will understand that:

- ☐ A mean is a measure of center.
- ☐ A mean can be computed for a population, a sample, and a set of samples.
- ☐ The Central Limit Theorem can be applied to approximate the mean of a population.

You will be able to:

- ☐ Determine the mean of a data set.
- ☐ Graph sample means and use the Central Limit Theorem to estimate the population mean.

2) In Lesson 1, Part A, we began collecting data for the number of cups of water that we drink each day. Let's analyze the plot of our data to help us identify the population mean, μ.

   Part A: What does each dot on the class dotplot represent?

   Part B: Describe what you consider to be the (a) center and (b) the overall distribution of the data. Does anything surprise you?

   Part C: How could we calculate the population mean of the data set, μ? How does the calculated population mean compare with your estimate of the center of data on the dotplot from Part B?

3) Compute your personal average (the mean of your data) water consumption, round to the nearest whole cup, and post the result on the class dotplot.

   Part A: What does each dot on this dotplot represent?

   Part B: Describe the (a) center and the (b) overall distribution of the data.

4) How does the daily data for the whole population (question 2) compare with the data for the student averages (question 3)?

We can use the data to make an inference about our average daily amount of drinking water.

5) On the plot of sample means, draw lines to indicate the amount that the article recommends for fluid consumption by men and by women. Compare the class average with the recommendations in the water article. Remember to pay attention to units!

Lesson 3, Part C How Much Water Does Our Class Drink?

Credit: kolotype/Fotolia

In the previous lesson, we read an article about water consumption and health, and reviewed the data about our daily water consumption, exploring sample and population means. For simplicity's sake, let's average together the recommendations for men and women and use a general recommendation of 11 cups (88 ounces) of fluid per day. In this lesson, we will compare this goal with our class average using a measurement called **standard deviation**.

1) Is our class average at, below, or above the goal of 11 cups of water per day?

Objectives for the lesson

You will understand that:

- ☐ Standard deviation is a measure of spread for continuous data.

You will be able to:

- ☐ Use standard deviation to interpret the spread of a data set.
- ☐ Calculate the percentage of data in a graph region.

You can calculate the standard deviation of a data set by using a spreadsheet or a free website such as http://easycalculation.com/statistics/standard-deviation.php

B5 | fx =AVERAGE(B2:B4)

	A	B	C	D
1		Mean Water Consumption (per day, in cups)		
2		5.2		
3		3.7		
4		8.5		
5	Mean	5.8		
6	St.Dev.			

B6 | fx =STDEV(B2:B4)

	A	B	C	D
1		Mean Water Consumption (per day, in cups)		
2		5.2		
3		3.7		
4		8.5		
5	Mean	5.8		
6	St.Dev.	2.455605832		

2) Using the averages for each student in the class, use a spreadsheet or a website to calculate the mean and standard deviation of the student averages.

Mean of averages =

Standard deviation of averages =

How many averages are in the data set (how many students brought their data)?

3) Calculate each of the following values:

Mean + 1(standard deviation) =

Mean + 2(standard deviation) =

Mean + 3(standard deviation) =

Mean – 1(standard deviation) =

Mean – 2(standard deviation) =

Mean – 3(standard deviation) =

4) Mark the mean and each of the values from question 3 on the sample averages dot plot.

Part A: Count the number of values between the ±1 standard deviation markers.

Part B: What percent of the values are between the ±1 standard deviation markers?

Part C: What percent of the values are between the ±2 standard deviation markers?

Part D: What percent of the values are between the ±3 standard deviation markers?

5) In the last lesson, you marked a line to show the water consumption recommendation from the article. Where does that recommendation lie in comparison to the standard deviation markers?

The class dotplot of student averages may or may not look like a normal distribution. However, when we calculate the standard deviation of large, normally distributed data set, we will find that:

Approximately 68% of the data values are within ±1 standard deviation of the mean.

Approximately 95% of the data values are within ±2 standard deviations of the mean.

Approximately 99.7% of the data values are within ±3 standard deviations of the mean.

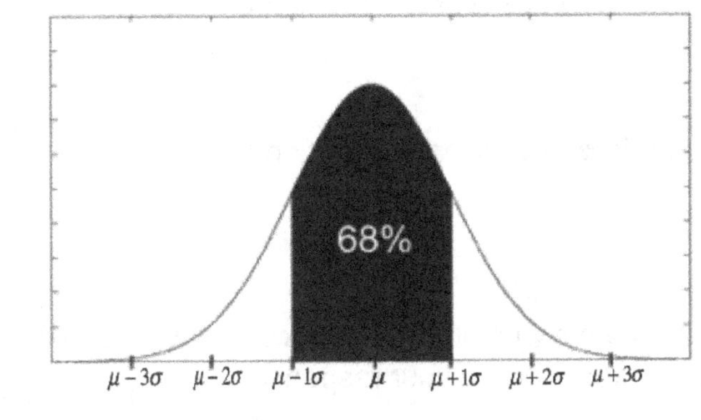

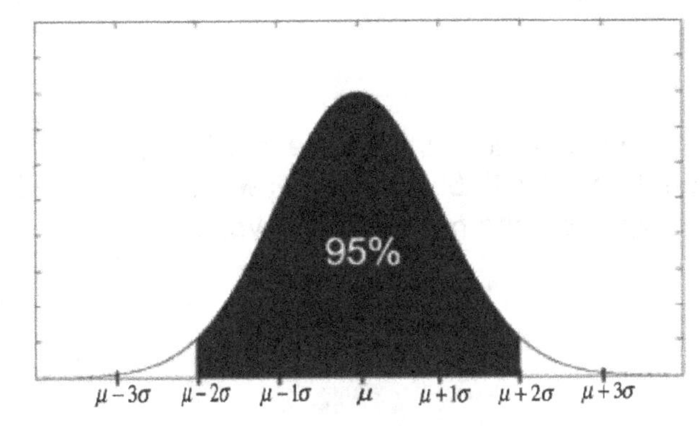

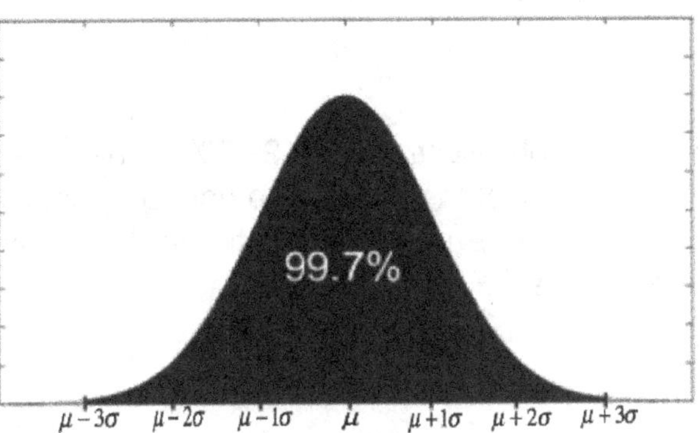

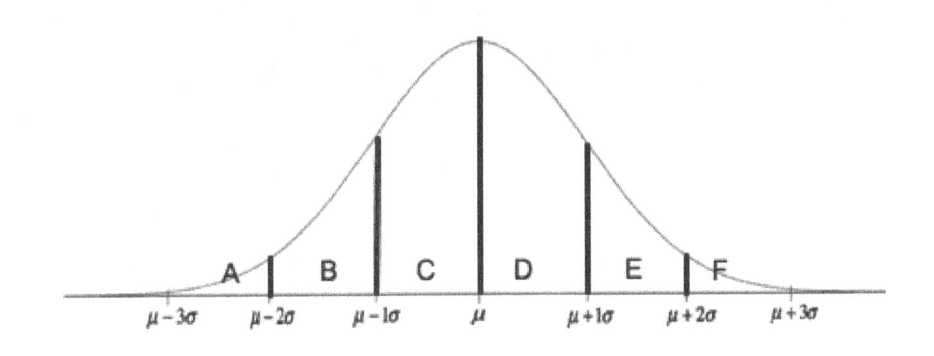

In your Preview Assignment, you were asked to print out the normal curve and place it in your notebook.

6) On your paper, draw vertical lines through each of the standard deviation markers. Mark the left-most region as A, the next region as B, etc.

   Part A: What percentage of data falls in Section C on the graph? How do you know?

   Part B: What percentage of data falls in Section D?

Part C: What percentage falls in Section B? Section E? How do you know?

Part D: What percentage falls in Section A? Section F?

7) In a normally distributed data set with 500 values, how many values will be within ±1 standard deviation of the mean? Justify your response.

8) The graphs illustrate that 99.7% of the data is expected to fall within ±3 standard deviations of the mean. In a normally distributed data set with 500 values, how many values will be more than 3 standard deviations away from the mean? Justify your response.

Lesson 4, Part A What Are the Risks?

In Preview Assignment 4.A, you studied some data from the American Cancer Society. You also read about using probability to better understand your risk.

Credit: Nolight/Fotolia

1) If you roll one die, what are the possible outcomes?

2) If you flip one coin, what are the possible outcomes?

Objectives for the lesson

You will understand:

- ☐ The concept of probability of independent events.
- ☐ How probability can help us to understand the risk of different situations.

You will be able to:

- ☐ Calculate theoretical probability of two or more independent events.
- ☐ Calculate "**and**" and "**or**" probabilities for independent events.

According to probability theory, two events are **independent** if the occurrence of one does not affect the probability of the occurrence of the other. For example, if you roll a die and get a 6, this result does not affect the probability of getting a 6 (or any other number) on your next roll.

Probabilities may be determined when two (or more) events occur at the same time (Event A <u>and</u> Event B) or when at least one of two (or more) events occur (Event A <u>or</u> Event B). Let's look at how the probability of independent "**and**" events and independent "**or**" events are computed.

3) If you roll one die, what is the probability of a 2? If you flip one coin, what is the probability of "heads"?

4) If you roll one die and flip one coin, does the occurrence of a particular outcome on the die affect the probability of a particular outcome on the coin? Why or why not?

5) If you roll one die and flip one coin, what are all the possible outcomes?

6) If you roll one die and flip one coin, what is the probability of rolling a 2 and flipping a head? Why?

7) Consider your answers to question 3 and question 6. How are the answers to each of these two independent events related to the probability of the events both occurring?

8) Refer to the website of information from the American Cancer Society: http://www.cancer.org/cancer/cancerbasics/lifetime-probability-of-developing-or-dying-from-cancer.

Part A: If we randomly select two U.S. females, what is the risk that they both will develop breast cancer sometime during their lifetimes? Assume that these are independent events. Explain your reasoning.

Part B: Is it possible that the choice of two females is not an independent event?

9) If you roll one die and flip one coin, what is the probability of rolling a 2 OR flipping a head? Why?

10) If you roll one die two times, what is the probability of getting a 2 on the first roll and a 2 on the second roll? Show work or explain your reasoning.

Independent "and" Events

To determine the probability of two independent events, A and B, **occurring together**:

P(A **and** B) = P(A) x P(B)

Independent "or" Events

To determine the probability of **one OR the other** of two independent events, A and B, occurring:

Add the probabilities of the events and subtract the intersection
P(A **or** B) = P(A) + P(B) – P(A and B)

Lesson 4, Part B Calculating Risk

The advances in cancer research have not eliminated the disease, but they have increased the probability of a longer life. This lesson helps you to be more critical about data when they are presented to you.

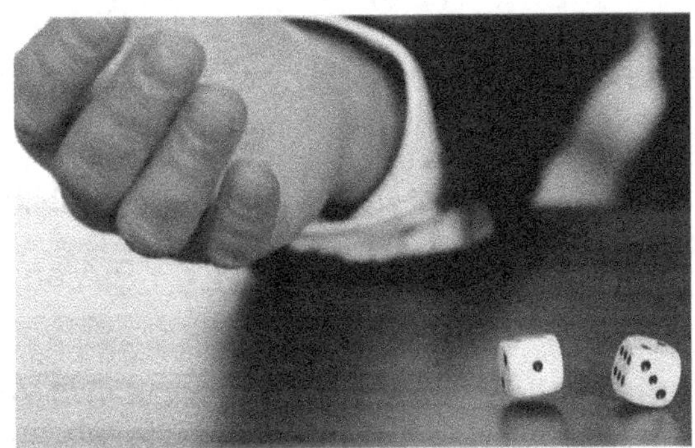

Credit: Piotr Marcinski/Fotolia

1) How do you think the mortality rate on the cancer website was calculated?

Objectives for the lesson

You will understand:

- ☐ The concept of conditional probability, including dependent events.
- ☐ How conditional probability can help us to understand the risk of different situations.

You will be able to:

- ☐ Calculate conditional probabilities for two or more dependent events.

Dependent Events

Events are considered to be **dependent** if the outcome of one event affects the outcome of another event.

Conditional Probability

Conditional probability measures the probability of an event given that another event has occurred. If the events are A and B, respectively, that is said to be "the probability of A given B." It is commonly denoted by P(A|B).

Conditional Probability of Dependent "and" Events

The conditional probability of Event A and Event B happening with dependent events is calculated by multiplying:

The probability of Event A by the probability of Event B given that A already happened.

P(A **and** B) = P(A) x P(B|A)

Example: A bag contains 4 yellow and 6 blue balls. The probability of drawing a yellow ball first **and** a yellow ball second [P(Yellow **and** Yellow)] is calculated by multiplying the probability that the first ball is yellow and the probability that the second ball is yellow given that a yellow ball has already been drawn.

The possible outcomes can be displayed as a tree diagram where the branches are labeled with the probability of each event.

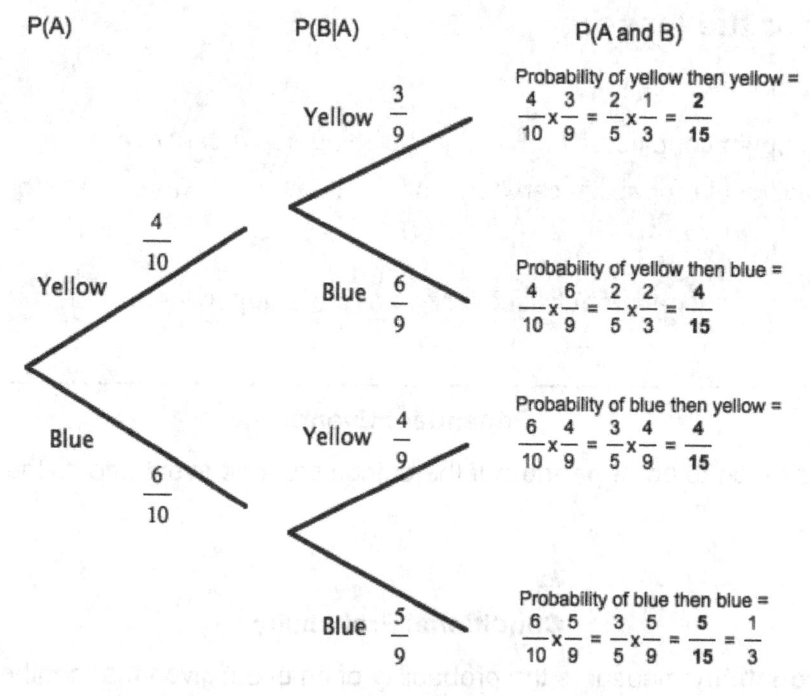

Note: The final probability of all the outcomes adds up to 1:

2/15 + 4/15 + 4/15 + 5/15 = 15/15 = 1

2) Let's do some rearranging.

Part A: If 3 x 4 = 12, what is $\frac{12}{3}$? If $(x)(y) = z$, what is $\frac{z}{x}$?

Part B: If P(A) x P(B|A) = P(A **and** B), what is $\frac{P(A \text{ and } B)}{P(A)}$?

Let's examine the American Cancer Society (ACS) data again from http://www.cancer.org/cancer/cancerbasics/lifetime-probability-of-developing-or-dying-from-cancer.

The article presents the risk for U.S. men of developing cancer ("All invasive sites") as 43.31% and the risk for U.S. men dying from cancer ("All invasive sites") as 22.83%. Let's take a closer look at these statistics. These numbers are based on incident and mortality data from 2009 to 2011.

- The incidence data for getting cancer were determined by dividing the number of U.S. males who have cancer by the number of U.S. males in the population.
- The mortality rate was calculated by dividing the number of U.S. male deaths due to cancer in one year by the total number of U.S. male deaths during that time period. Look at it another way: Given all the male deaths in one year, what percent were due to cancer?

The mortality rate does not tell us what the chances are if you are male and diagnosed with cancer or what your risk is for dying from that cancer. In other words, the two sections of the table have different base numbers. Let's translate this information into probability notation.

In questions 3–9, round answers to the nearest percent.

3) What is the probability that a U.S. male will have cancer in his lifetime?

   P(will have cancer | U.S. male) = ___________

4) What is the probability that a U.S. male will die from cancer?

   P(die from cancer | U.S. male) = ___________

5) What is the probability that a U.S. male will die from cancer if he has been diagnosed with cancer?

   P(male die from cancer | male has cancer) = P(die from cancer) / P(has cancer) =

   ____

Use the process described to determine the following conditional probabilities.

6) What is the probability that a U.S. female will die from breast cancer if she is diagnosed with breast cancer?
   P(female die from breast cancer | female has breast cancer) = ___________

7) What is the probability that a U.S. female will die from lung cancer if she is diagnosed with lung cancer?
   P(female die from lung cancer | female has lung cancer) = ___________

8) What is the probability that a U.S. male will die from lung cancer if he is diagnosed with lung cancer?
   P(male dying from lung cancer | male will have lung cancer) = ___________

9) Do males or females have a higher risk of dying from melanoma of the skin, knowing that they have the melanoma? Show your solution process and explain your results.

10) Write a statement comparing and contrasting the independent and dependent events for the probability of dying from melanoma.

Lesson 5, Part A Cost of Living Comparisons

You have been offered three full-time jobs, each in different cities and with different salaries. Other than salary and city, the job responsibilities are exactly the same.

- The first job offers an annual salary of $65,000 and is in Austin, TX.
- The second job offers an annual salary of $60,000 and is in Big Rapids, MI.
- The third job offers an annual salary of $100,000 and is in Thousand Oaks, CA.

Credit: xixinxing/Fotolia

1) How would you decide which job to accept?

Objectives for the lesson

You will understand that:

- ☐ A mathematical tool is needed in order to compare monetary values.
- ☐ Dollars have different values and purchasing power in different cities.

You will be able to:

- ☐ Recognize when converting units is needed.
- ☐ Use conversions to make comparisons.

In this lesson, we focus on the purchasing power of the dollars in each city, rather than other factors (such as climate, proximity to family, etc.) that could impact the decision about accepting the job.

2) Consult the spreadsheet containing average annual costs for essential goods and services (economists call this a "basket of goods") for Austin, Big Rapids, and Thousand Oaks. The sum of these items (e.g., housing, utilities, food, transportation) is often called the "cost of living" in the city. Using this data, determine how much it would cost to live per year in each of these three cities.

3) Write the ratios for the annual cost of living for each pair of cities and reduce each ratio to a unit ratio. (Round to two decimal places.)

Austin to Big Rapids

Big Rapids to Austin

Thousand Oaks to Austin

Austin to Thousand Oaks

Big Rapids to Thousand Oaks

Thousand Oaks to Big Rapids

4) We can think of dollars in different cities as having different buying power in different cities. How would you use your unit ratios from question 3 to express the buying power of the dollars in one city to the dollars in another city?

To convert Austin dollars to Big Rapids dollars

To convert Big Rapids dollars to Austin dollars

To convert Thousand Oaks dollars to Austin dollars

To convert Austin dollars to Thousand Oaks dollars

To convert Big Rapids dollars to Thousand Oaks dollars

To convert Thousand Oaks dollars to Big Rapids dollars

5) Use your conversion factors from question 4 to determine the value of each salary in each of the three cities. Fill in the table below with your responses.

	Austin Salary	**Big Rapids Salary**	**Thousand Oaks Salary**
Value in Austin	$65,000		
Value in Big Rapids		$60,000	
Value in Thousand Oaks			$100,000

6) Write a brief interpretation of each number you calculated in question 5.

7) Use your response to question 6 to select the job that you will accept and write a brief one-sentence justification that will convince your family.

Lesson 5, Part B Index Numbers

1) If you have access to technology, open the spreadsheet you used in Preview Assignment 5.B that listed the salaries and costs of living in ten cities. Review the data that you added in Column E. Write one sentence explaining the practical meaning of a percentage in Column E that is less than 100% or more than 100%.

Credit: Barabas Attila/Fotolia

Objectives for the lesson

You will understand:

- ☐ The meaning of index numbers.
- ☐ The application of index numbers to your life.

You will be able to:

- ☐ Perform calculations involving index numbers.
- ☐ Make and justify decisions and evaluate claims using index numbers.

2) Before using our data from the spreadsheet for new calculations, let's make sure that we can make sense of certain calculations that we will need to do.

Suppose that $\frac{A}{B} = 0.59$. Which of the following calculations is equivalent to 143 x $\frac{B}{A}$?

Check all that apply.

a) 143×0.59

b) $143 \div 0.59$

c) $\frac{143}{0.59}$

d) $143 \times \frac{1}{0.59}$

e) $\frac{1}{143} \times 0.59$

3) Refer to your spreadsheet of the salaries and costs of living in ten cities. We will use the national average to convert each city's salary to an equivalent salary based on the national average cost of living. Then we can compare all ten salaries.

In order to perform the conversion, we must change units. For example, we need to change Phoenix dollars to national average dollars.

The percentages you performed in the Preview Assignment were obtained by dividing Phoenix dollars by national dollars. However, to convert units, we need to calculate:

$$National\ Equivalent\ Salary = Phoenix\ Salary \times \frac{National\ Cost\ of\ Living}{Phoenix\ Cost\ of\ Living}$$

Perform this calculation.

4) Describe in one or more complete sentences how units are involved in question 3.

5) Describe in one or more complete sentences how the formula described in question 3 is similar to converting units.

The percentages that you calculated in Column E of the spreadsheet are called the **cost of living index** for each city. This is an example of an **index number** (used for comparisons). Index numbers are numerical tools used to compare measurements made in different places or in different times. The plural of index is **indices**.

6) In the last lesson, you used cost of living ratios to compare three job offers. Now we are going to use the cost of living index to compare several job offers.

   Think of the cost of living index as a unit conversion, which converts the dollars in one city to dollars in "Anytown, USA," given the national average cost of living.

$$National\ Equivalent\ Salary = City\ Salary\ x \frac{National\ Cost\ of\ Living}{City\ Cost\ of\ Living}$$

   Use the concepts given and your spreadsheet to convert each salary to equivalent dollars in Anytown, USA. Report the result of the calculations in Column F of the spreadsheet.

7) Decide which job that you will select and write a one-sentence justification that would convince your family.

8) The Cost of Living Index helps us measure how the value of a dollar varies across the country. The value of the dollar has to do with where in the country we are living. Besides by location, what else might cause the value of a dollar vary?

You may have encountered index numbers before, such as Body Mass Index (BMI) used to measure obesity and the Consumer Price Index (CPI) used to measure inflation. Common examples (emphasized in this lesson) are the cost of living index (which allows for comparisons of the value of money in different locations), body mass index (which measures obesity), and the consumer price index (which allows for comparisons of the value of money over time accounting for inflation).

Lesson 5, Part C Polls, Polls, Polls!

Pretend it is almost election time in the year 2144. You are on the reelection staff of President Kathleen Smith. It is October 21, and election day (November 3) is approaching quickly.

Your job is to analyze the results of polls conducted by three different companies and report to President Smith.

Poll	Dates	Sample Size	Result (Pro-Smith)
Poll A	10/4 – 10/6	1024	45%
Poll B	10/18 – 10/20	2048	51%
Poll C	10/18 – 10/20	1596	52%

1) After looking at the poll results, what information would you give President Smith about her chances of reelection? Which of the three polls should be given the most weight?

Objectives for the lesson

You will understand:

- ☐ What a weighted average is and how it is used.

You will be able to:

- ☐ Calculate weighted averages.
- ☐ Use weighted averages to analyze data and draw conclusions about the data.

2) Disregarding sample size, calculate the average (mean) of the results from the three polls. (Round to the nearest percent.) Based on this result, what do you tell the President?

3) Let's consider which of the three polls are least and most reliable (and therefore likely to be less or more important to the President). Reliability is a term used to communicate whether a poll (or survey) could be reproduced. In other words, reliability is a measure of precision.

Part A: Of the three polls, which may be the <u>least</u> reliable? Why? How is that poll affecting your average from question 2?

Part B: Which poll(s) may be the most reliable? Why? How might that affect your average from question 2?

To make a statistical correction for reliability, polls are usually combined using **weighted averages**. Weighted averages are averages that are computed after assigning each quantity a weight that identifies its relative importance. Each quantity is multiplied by its respective weight and the results are added. It is important that the weights total to 1.0 (or 100%).

The weighted average is calculated using the following two-step procedure:

Step 1: Multiply the weight by the result for each poll. (Do not convert the poll percentage to a decimal; just multiply by the number.)

Step 2: Add the products from Step 1. The sum is the weighted average.

The selection of weights is normally the result of advanced scientific analysis.

4) In our Presidential poll, we will make a good guess about the weights based on the information about each poll. Let's assign Poll 1 a weight of 0.2 (or 20%) and Polls 2 and 3 each a weight of 0.4 (40%).

The polls and weights are summarized in the new table shown. Use the two-step procedure to calculate the weighted average of the three polls (using the given weights).

Poll	Result (pro-Smith)	Weight	Result x Weight
Poll 1	45%	0.2	
Poll 2	51%	0.4	
Poll 3	52%	0.4	
		Total = 1.0	Weighted Average:

5) Compare your results in questions 2 and 4.

Part A: Does the advice that you provide to the President change if you use a weighted average?

Part B: Which should be more accurate?

Part C: Do you think the weights should be adjusted?

6) The spreadsheet for this lesson contains data from fifteen polls along with the weight of each poll. Use the spreadsheet to calculate the weighted average and use this information to advise the President. Report the weighted average in Cell D17, save the spreadsheet to your computer, and record your result. How is the President doing in the polls?

7) Which of the polls in the spreadsheet do you think are most recent or have the largest sample size? Why? How does that affect your final answer?

Lesson 5, Part D Average Income

In the Preview Assignment, you calculated the simple overall average salary of the ten cities shown.

1) Multiply the population of Phoenix times the average salary in Phoenix. What does the answer represent?

City	2010 Census Population	Salary
Phoenix, AZ	1,488,750	$98,000.00
San Francisco, CA	825,863	$125,000.00
Denver, CO	634,265	$101,000.00
Detroit, MI	701,475	$78,000.00
Crete, NE	7,174	$69,000.00
Schenectady, NY	154,727	$95,000.00
Amarillo, TX	195,250	$80,000.00
Dallas, TX	1,241,162	$102,000.00
El Paso, TX	672,538	$80,000.00
San Antonio, TX	1,382,951	$79,000.00
Total	7,304,155	

Objectives for the lesson

You will understand:

- ☐ The meaning of expected value.
- ☐ That an expected value is a weighted average.

You will be able to:

- ☐ Calculate expected value
- ☐ Make predictions about real-world scenarios based on your knowledge of averages, weighted averages and expected values.

2) How would you calculate the weighted average of the salaries?

In this lesson, we extend our discussion of weighted average by examining something called "expected value."

Expected Value

Expected value is a weighted average in which we work with percentages rather than raw data. We determine a value's percentage of the total and then multiply that percentage by the variable of interest to calculate the expected value.

Let's apply these principles to compare the salaries and populations of the ten cities listed in the table. Open your spreadsheet or locate the population percentages that you calculated in Preview Assignment 5.D.

3) Calculate the amount of money that each city contributes to the expected value of the overall average salary.

City	2010 Census Population	Percentage of Total Population	Salary
Phoenix, AZ	1,488,750		$98,000.00
San Francisco, CA	825,863		$125,000.00
Denver, CO	634,265		$101,000.00
Detroit, MI	701,475		$78,000.00
Crete, NE	7,174		$69,000.00
Schenectady, NY	154,727		$95,000.00
Amarillo, TX	195,250		$80,000.00
Dallas, TX	1,241,162		$102,000.00
El Paso, TX	672,538		$80,000.00
San Antonio, TX	1,382,951		$79,000.00
Total	7,304,155	100%	

4) Now find the total of the ten cities' contributions. The result is the expected value of average salary.

5) Describe the similarities and differences between expected value and weighted average. Why do you think knowing the relationship between the two calculations is helpful?

Lesson 6, Part A How Can We Smooth the Data?

In the Preview Assignment, you studied data that was erratic and had a jagged graph. You learned about simple moving averages that sometimes can help to smooth the data and help us see the overall trend.

1) Louis' quiz scores and the corresponding moving averages are shown in the graph from Preview Assignment 6.A. What conclusions did you make about his performance on quizzes after looking at the two graphs?

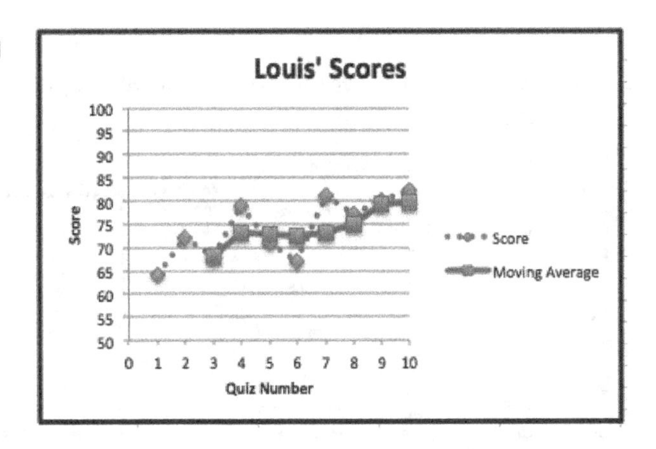

Objectives for the lesson

You will understand:

- ☐ The reason for weighted moving averages.

You will be able to:

- ☐ Calculate simple and weighted moving averages.
- ☐ Analyze graphs of moving average data.

Another method for smoothing an erratic line graph and to help us determine future trends is called a **weighted moving average**. This type of average is commonly used to analyze stocks and to help predict market trends.

What is a weighted (exponential) moving average?

Similar to a simple moving average, a weighted moving average takes groups of values, shifting down one value at a time and computing the average for each group, until all of the groups of values have been averaged. In a weighted average, each value in the group is given a different weight, with the most recent value having the heaviest weighting.

A common strategy is to group three data points and assign weights of 1, 2, and 3, respectively. The most recent data value in the group is assigned a weight of 3 and the oldest data value is assigned a weight of 1.

Calculations: Each of the three data values is multiplied by its weight then the three products are added together. This sum is divided by 6, since the sum of the weights is 6 (1 + 2 + 3).

2) Let's calculate the weighted moving average using Louis' quiz scores from the Preview Assignment.

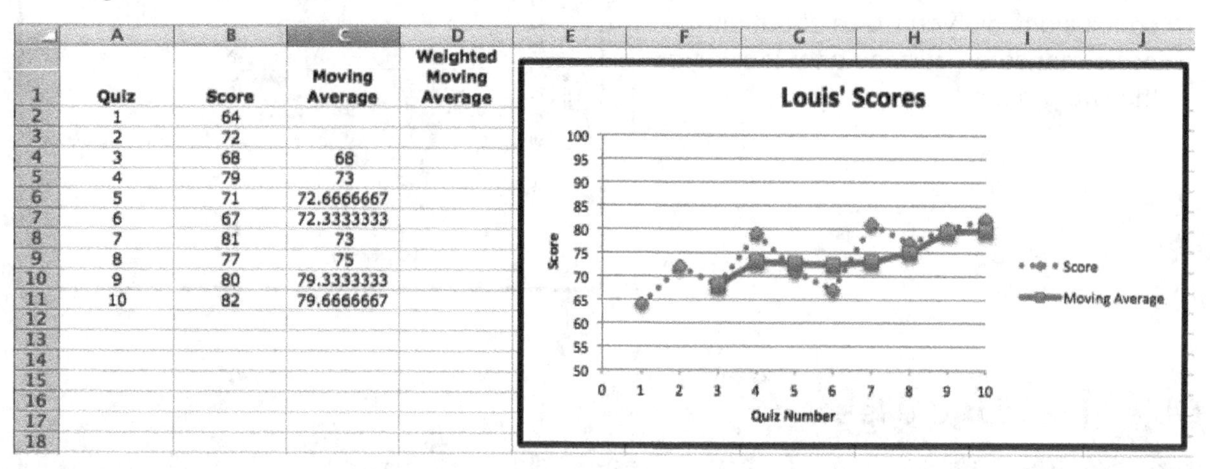

Quiz	Score	Moving Average	Weighted Moving Average
1	64		
2	72		
3	68	68	
4	79	73	
5	71	72.6666667	
6	67	72.3333333	
7	81	73	
8	77	75	
9	80	79.3333333	
10	82	79.6666667	

Part A: Find D4 = (1*B2 + 2*B3 + 3*B4)/6 = __________. Round to the tenths place, if needed.

Part B: Continue this process to find the remaining values in Column D. Then plot the weighted moving average on the graph.

Part C: What do you notice about the relationship between the three graphs?

Part D: What conclusion can you draw from the trend of Louis' scores?

3) Describe the difference between a simple moving average and a weighted moving average. Why might you choose to do a weighted moving average?

4) How would you modify this process if you wanted to use <u>four</u> data values?

Lesson 6, Part B Mini Project: Income Disparities

In Practice Assignment 6.A, we analyzed the historical trend for the Top 5 Percent and the Lowest Fifth (lowest 20%) mean household income levels from U.S. Census Bureau data.

In this mini-project, you will determine the trend for the remaining categories and write a report about your findings.

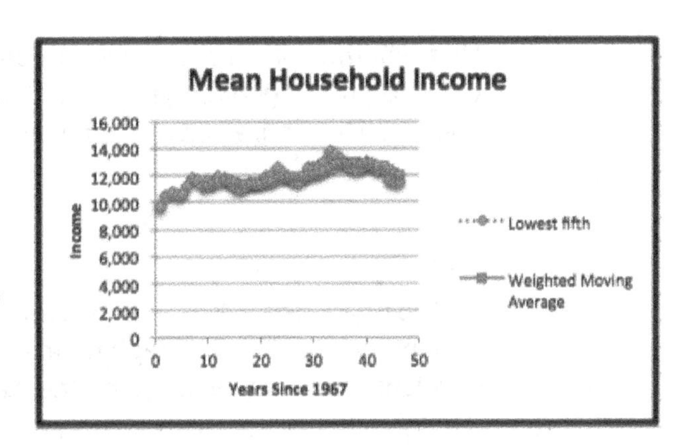

1) Do you predict that the middle class income in 2012 was more than, the same, or less than in 1967 (in 2012 adjusted dollars)?

Objectives for the lesson

You will understand:

- ☐ The concept of weighted (exponential) moving averages (EMA) and why the method is used.

You will be able to:

- ☐ Calculate and compare simple and weighted moving averages.
- ☐ Write a contextual analysis of a graphical display of weighted average data in a formal paper (at least two paragraphs long), including appropriate mathematical language and explanations.

We will use the same process—calculating weighted moving averages—to show basic trends for the Second Fifth (lower middle-class), Third Fifth (middle middle-class), and Fourth Fifth (upper middle-class) mean household incomes of the U.S. federal budget.

We will also calculate the weighted moving averages of the mean income levels of the upper class (the Highest Fifth). Once we have determined the trend, we will compare the trends of the mean household income levels of the middle classes with upper and lower income classes.

2) Open Spreadsheet 6.B and click on the **Lesson 6, Part B Mini-Project** tab.

Part A: In four open columns to the right of the data (Columns I, J, K, and L), calculate the weighted moving average (EMA) using groups of ten data values for the Second, Third, Fourth, and Highest categories (Columns C, D, E, and F). Record your averages in the four open columns. Prepare your data for submission according to your instructor's directions (e.g., save and submit the spreadsheet or make a printout of the weighted moving average columns.)

Hint: Put the columns right next to each other. Create a formula for the first weighted moving average box, and then copy and paste it to fill in the rest of the column. This entire column can now be copied to the other columns.

Now you are going to create four line graphs. Make one graph for each of the four categories mentioned in Part A, graphing the actual data from 1967 through 2012 and the respective weighted moving average on the same grid for each of the four categories.

Part B: Create the graph comparing the actual data to the weighted moving average data for the Second Fifth. Then write a short description.

Hint: To make the graphs, highlight only the columns you are interested in including on each graph. To highlight, use the Command key on a Mac computer or the Alt key on a PC.

3) Now consider all four graphs together. What can you conclude about the trends of all four categories?

4) The U.S. Census Bureau reports that the median household income in 2012 was $51,000. What is a median and what does it represent in this context? Write a complete contextual response.

Lesson 7, Part A U.S. Budget Priorities

The categories of a financial budget indicate an individual's or group's expense items. By analyzing the amount of money in each line item, you may be able to draw conclusions about priorities. Personal budgets can include line items such as housing (rent or mortgage), utilities, food, clothing, entertainment, and education.

1) In the Preview Assignment, you made a guess about the top three categories in the federal budget.

   Part A: If it were your choice, which three categories would you prioritize?

   Part B: Write down the percentage of the overall budget that you think each of your top three categories should represent, and explain why you made these choices.

Federal Budget Categories

- Education
- Energy and Environment
- Food and Agriculture
- Government
- Housing and Community
- Interest on Debt
- International Affairs
- Medicare and Health
- Military
- Science
- Social Security, Unemployment, and Labor
- Transportation
- Veterans Benefits

Objectives for the lesson

You will understand that:

- ☐ Ratios can be represented by fractions or percentages.
- ☐ Percentages can be used to represent part-to-whole ratios.

You will be able to:

- ☐ Determine percentages based on part-to-whole ratios.
- ☐ Write a ratio or percentage and explain its meaning within a context.
- ☐ Read a budget, determine values of line items, and draw conclusions about the overall distribution of funds.

Now that you have listed your top three categories, let's look at the actual values in all of the categories of the proposed U.S. federal budget. The pie graph shown represents the 2014 federal budget that was proposed by President Obama.

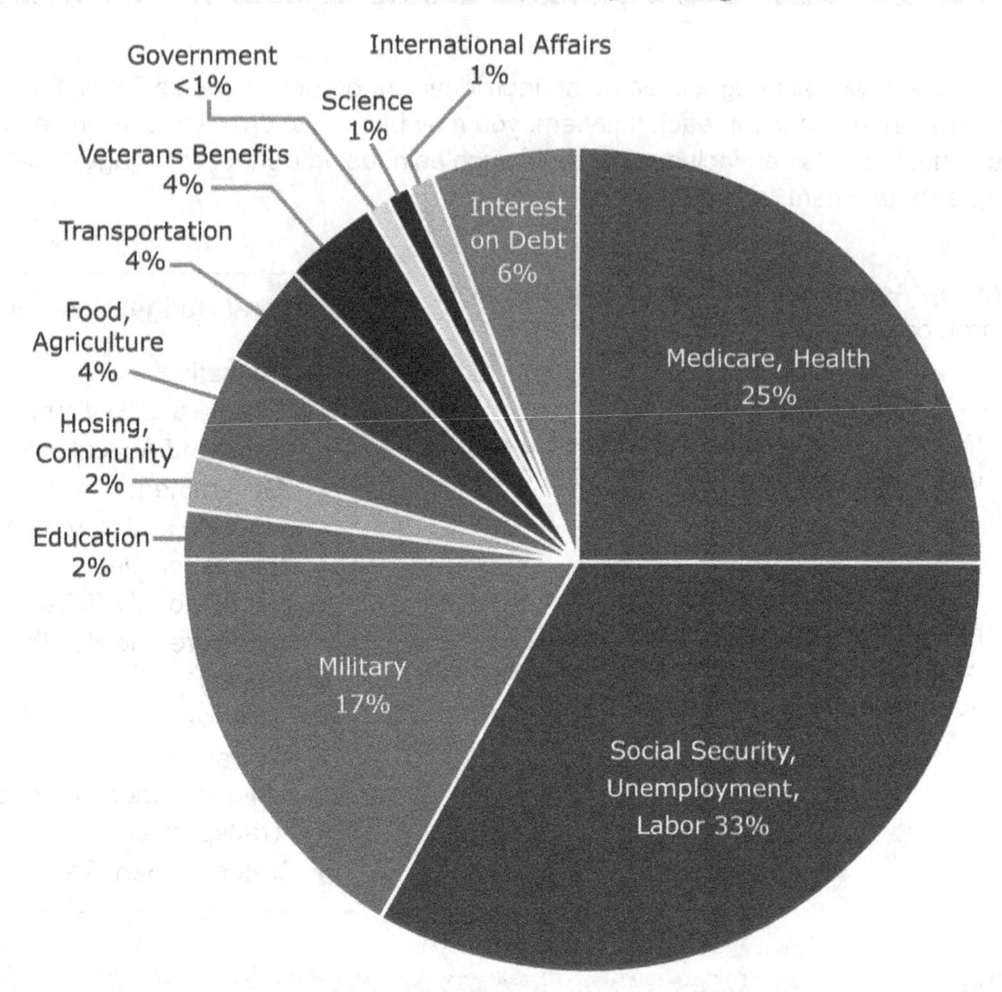

Data from: https://www.nationalpriorities.org/analysis/2013/president-obamas-fiscal-year-2014-budget/

2) Identify the actual top three budget categories shown in the pie graph.

3) How do the categories on the pie graph align with what you predicted were the actual top three categories from Preview Assignment 7.A?

4) The total U.S. federal budget for 2014 was $3,777,807,000,000. Using the pie graph percentages, calculate the amount of money budgeted, rounded to the nearest billion, for the top three categories listed on the pie graph.

Part A: Social Security, Unemployment and Labor = ____________________

Part B: Medicare and Health = ____________________

Part C: Military = ____________________

5) In question 4, you calculated how much each of the three categories represented out of the total budget. Now let's compare two budget categories.

   Part A: Write a ratio of the amount budgeted for the military compared with the amount budgeted for the Social Security, Unemployment, and Labor category.

   Part B: What common fraction is this close to?

   Part C: Write a contextual sentence explaining what the answer to Part B means.

   Part D: Convert the ratio from Part A or B to a percentage.

   Part E: What does the percentage mean in this situation?

6) What are your observations about the distribution of funds into the categories of the U.S. federal budget? What do you conclude about the priorities of government spending?

Lesson 7, Part B Understanding U.S. Budget Priorities

In Preview Assignment 7.B, Megan's time budget for a 24-hour day showed that she puts more time into her schoolwork than into her current job. Her budget may reflect a higher value on preparation for the future than on short-term needs.

Budgets say something about our values. The more important items receive more resources than less important items.

Credit: zimmytws/Fotolia

1) What is a budget? How can you use a budget? What do percentages of money spent for different items in a budget tell you about priorities?

Objectives for the lesson

You will understand:

- ☐ The meaning and use of part-to-part and part-to-whole ratios.
- ☐ That part-to-whole ratios can be used to create a pie graph.
- ☐ That the meaning of percentages depends on the context.

You will be able to:

- ☐ Use part-to-part ratios, part-to-whole ratios, and percentages to calculate ratios and compare line items in budgets.
- ☐ Use ratios and percentages to construct a pie graph.
- ☐ Examine and interpret ratios, percentages, and pie graphs.

The spreadsheet for this lesson contains the 2012 U.S. federal budget.[1] Each row is a **line item**—a spending category. The data tell you how much money (in millions of dollars) was spent on each line item in 2012. In the Preview Assignment, you were asked to look up the meaning of each line item. Now we will analyze and compare the amount of money allocated in each line item.

---

[1] Source: Table 3.1—Outlays by Super function and Function: 1940–2018. Retrieved September 10, 2014, from http://www.whitehouse.gov/omb/budget/historicals.

	Federal Budget: 2012	2012
1	**Federal Budget: 2012**	**2012**
2		
3	Line Item	In millions of dollars
4	National Defense	677,852
5	Human Resources	2,348,587
6	Physical Resources	215,463
7	Net Interest	220,408
8	Other Functions	178,353
9	Undistributed Offsetting Receipts	-103,536
10	Total	3,537,127
11		
12		
13	Source: http://www.whitehouse.gov/omb/budget/historicals	Table 3.1 for the year 2012
14		

2) Write down the following ratios using the 2012 federal budget data.

Part A: National Defense spending to Total federal budget.

Part B: National Defense spending to Human Resources spending.

3) Compare the two ratios. How are they similar? How are they different?

4) There are two types of ratios: **part-to-part ratios** and **part-to-whole ratios**. For the two ratios you wrote down in question 2, which is a part-to-part ratio and which is a part-to-whole ratio? Justify your answer. What do you think is the difference?

5) What do these ratios tell you about federal budget priorities? Which of these ratios helps you to understand federal budget priorities better? Why?

6) For the ratio that you did not find as helpful for question 5, what additional information would make this ratio more useful?

7) In question 2, you wrote two ratios:

   Ratio A: National Defense spending to Total federal budget.
   Ratio B: National Defense spending to Human Resources spending.

   Which of these ratios would make the most sense if converted to a percentage? Why?

8) Calculate the percentages for the ratios in question 2. Round each to the nearest whole percentage point. Write contextual sentences to interpret the percentages.

   Part A: National Defense spending to Total federal budget.

   Part B: National Defense spending to Human Resources spending.

9) For each line item in the spreadsheet, create a part-to-whole ratio and calculate the percentage for the ratio. Round to the nearest whole percentage point. Be sure the sum of the percentages is 100%.

   Human Resources:

   Physical Resources:

   Net Interest:

   Other Functions:

   Undistributed Offsetting Receipts:

10) Consider your answers to questions 8 and 9.

   Part A: Create a pie graph of the part-to-whole percentages.

   Part B: Where is the most spending?

   Part C: Where is the least spending?

11) What do your answers to questions 9 and 10 tell you about government priorities in 2012? Did the pie graph help you to analyze priorities? Why or why not?

Lesson 7, Part C Changes to U.S. Budget Priorities

Just as our personal budget priorities change over time, the federal government's priorities are also adjusted based upon the needs of the nation. In your Preview Assignment, you analyzed changes in the Education, National Defense, and Net Interest budgets.

1) Without looking at the spreadsheet, name another government budget priority that you think has changed (increased or decreased) over time?

Credit: niyazz/Fotolia

Objectives for the lesson

You will understand that:

- ☐ A line graph can model a trend over time.
- ☐ Hypotheses need to be supported by evidence.

You will be able to:

- ☐ Analyze data in a spreadsheet and graphs, using additive (absolute) comparison and multiplicative (relative) reasoning.
- ☐ Develop a reasonable hypothesis supported by evidence.
- ☐ Use spreadsheets to create a line graph and describe the pattern of the graph.

2) Examine the spreadsheet data from Practice Assignment 7.C. The figures in the spreadsheet are in millions of dollars.

Part A: What factor of 10 must be multiplied by each data value to show all the decimal places in the number?

Part B: In 2009, the amount budgeted to Education was $79,749 million. Write this number with the correct number of zeros. What is another way to write this number?

3) Consider the data for the line items Education, National Defense, and Net Interest over the years 2000 and 2014. Fill in the table cells with the appropriate budget amount.

	2000	2014
Education, Training, Employment, and Social Services		
National Defense		
Net Interest		

4) Let's focus on the Education budget.

Part A: What is the absolute change in the amount of money allocated to education from 2000 to 2014? (Absolute change is also called additive reasoning.)

Part B: Calculate the relative change in the amount allocated to education between 2000 and 2014. (Recall that relative change is calculated by taking the difference between the new and older amounts and then dividing by the older amount. The answer is then converted to a percentage. Relative change is also called multiplicative reasoning.)

Part C: What is your hypothesis about the federal government's priority for education? Use your calculations and your graph to support your hypothesis.

5) Now choose either National Defense or Net Interest for the 2000–2014 period. Perform the same analysis as you did in question 4. Write a brief paragraph about your chosen category to be read on the morning news show. Refer to your calculations and/or your graph.

Lesson 7, Part D Percent of Total U.S. Budget

In the previous activity, you discovered that the education category in the federal budget had increased by more than $75 billion over the previous 14 years.

1) What are some reasons that the Education budget might have been increased?

Credit: kg2cwow7hm5/Fotolia

Objectives for the lesson

You will understand that:

- ☐ Claims and hypotheses may be adjusted based on new evidence.

You will be able to:

- ☐ Analyze data in spreadsheets and graphs to compare changes in categories.
- ☐ Revise a claim or hypothesis based on new evidence.

2) In Lesson 7, Part C, you calculated the absolute change and the relative change in the Education, National Defense, and Net Interest categories. Now let's do the same for the Total, Federal Outlays category. Find the table from Lesson 7, Part C, in your notes. Add a new row for Total, Federal Outlays. Then complete the questions shown.

Part A: Calculate the absolute change from 2000 to 2014 in Total, Federal Outlays.

Part B: Calculate the relative change from 2000 to 2014 in Total, Federal Outlays.

Part C: Compare your answer to Part B to the relative change in the Education category. What, if anything, does this tell you about changes to the U.S. budget priorities?

3) Create a new row in the spreadsheet that shows the percentage of the total budget that is allocated to Education.

Part A: What percentage of the federal budget was allocated to Education in 2000? In 2014?

Part B: Create a graph of this new data, **Education as a Percentage of Total Federal Outlays**.

Part C: What are some interesting features of this graph? What does it tell you about U.S. federal budget priorities? Use evidence from the graph, the data, and the historical timeline to support your comments.

4) Perform a similar analysis for the National Defense category. Create a new row in the spreadsheet that shows the percentage of the total budget that is allocated to National Defense.

Part A: What percentage of the federal budget was allocated to defense in 2000? In 2014?

Part B: Create a graph of this new data, **National Defense as a Percentage of Total Federal Outlays**.

Part C: What are some interesting features of this graph? What does it tell you about U.S. federal budget priorities? Use evidence from the graph, the data, and the historical timeline to support your comments.

Lesson 7, Part E What's My Credit Score?

Recall that the FICO score is used to determine how much money you can borrow and how much interest you'll pay.

Credit: danielfela/Fotolia

1) Which of the statements below would be most likely to affect your credit score in a positive way? Why?
   - □) I pay everything with cash.
   - □) I make enough money to offset my debt.
   - □) I spread out my borrowing on a lot of cards, so I don't have one card with a huge balance.
   - □) I only have one credit card.
   - □) I only have student loans and one store credit card.

Objectives for the lesson

You will understand:

- ☐ Debt-to-income (DTI) ratios and the effect of those ratios on your credit score and ability to manage debt.

You will be able to:

- ☐ Calculate a DTI ratio.
- ☐ Draw a conclusion from the DTI about the appropriateness of the percentage of income spent on housing and debt.

In Preview Assignment 7.E, you were introduced to FICO credit scores and to debt-to-income (DTI) ratios. In this lesson, we will focus on taking out and paying back loans. We will begin with the DTI ratios used to determine a person's ability to manage and pay back debt.

2) Jot down what you remember about debt-to-income ratios from the Preview Assignment.

Debt-to-income ratios are used to determine a person's ability to obtain a mortgage. Most lenders examine Front-End DTI and Back-End DTI to determine a buyer's suitability for a mortgage loan.

Front-End DTI = $\dfrac{\textit{Housing payments}}{\textit{Total income}}$. Housing refers to rent or mortgage payments.

Back-End DTI = $\dfrac{\textit{All recurring debt payments}}{\textit{Total income}}$. Recurring debts include rent or mortgage, plus car payments, student loans, credit card payments, etc.

According to Bankrate, your Front-End DTI should not exceed 28% and your Back-End DTI should not exceed 36%.[1]

3) Suppose you want to move out of your apartment and buy a small house so that you can claim a deduction for the mortgage interest on your tax return. Let's assume that your income is $39,600 per year, your student loans cost $250 per month and your car payment is $325 per month. You only use your credit card for gas and pay off about $100 monthly. You used an online loan payment calculator and determined that the house payment will be $725 per month. Round answers to the nearest whole percentage point, where necessary.

Part A: What is your Front-End DTI ratio?

Part B: What is your Back-End DTI ratio?

Part C: Is the bank likely to lend you the money to buy the house?

Part D: What should your total monthly payments be to keep you in line with the Back-End DTI limit stated by Bankrate.com?

Part E: How much lower should your house payment be to stay within this limit?

[1] Source: http://www.bankrate.com/finance/mortgages/why-debt-to-income-matters-in-mortgages-1.aspx

Use one of the free online loan calculators listed below—or search one of your own—to answer questions 4 and 5.

http://www.myfico.com/myfico/creditcentral/loanrates.aspx

http://www.nytimes.com/interactive/2014/your-money/student-loan-repayment-calculator.html?_r=4

https://studentloans.gov/myDirectLoan/mobile/repayment/repaymentEstimator.action

http://partners.leadfusion.com/tools/myfico/budget05/tool.fcs?param=tWN*rVRxpGkNqmJ5tW1xrQ@@

4) After four years of college, let's assume that you owe $30,000 in student loans that charge 3.8% interest. How long will it take to pay off these loans (in years) if you pay $200 per month?

5) If, after four years of college, you owe $30,000 in student loans that charge 3.8% interest, how much would you have to pay monthly to pay off the loan in five years?

Lesson 7, Part F U.S. Incarceration Rates

In the article for the Preview Assignment, you read about aspects of the American justice system that are actually unique in the world.

Credit: mik38/Fotolia

At the time of this article, the United States had 2.3 million criminals behind bars, more than any other nation, according to data maintained by the International Center for Prison Studies at King's College London.

1) The U.S. does not have the largest population in the world. List a few of the factors you think might contribute to the U.S. having the most prisoners.

Objectives for the lesson

You will understand:

- ☐ Proportional reasoning, by writing and examining ratios, rates, and percentages.
- ☐ Graphical displays, and associated data.

You will be able to:

- ☐ Interpret ratios and percentages as rates of change.
- ☐ Compare two or more ratios and percentages.
- ☐ Read and interpret graphical displays.

Let's consider additional information and data posted by the American Civil Liberties Union (ACLU) on their "Safe Communities" blog.[1]

- The U.S. has 5% of the world's population and more than 20% of the world's prison population.
- Since 1970, our prison population has risen 408%.
- One in 110 adults in the U.S. are living behind bars.

---

[1] American Civil Liberties Union. (n.d.). The Prison Crisis | American Civil Liberties Union. Retrieved October 1, 2015, from https://www.aclu.org/safe-communities-fair-sentences/prison-crisis

- Counting prison, jail, parole, and probation, one in 35 adults in the U.S. are under some form of correctional control

Let's look closely at the statistics given in each source and compare and contrast the information.

2) In October of 2015, the estimated U.S population was 321.9 million and the world was 7,276.3 million.[2] Using these numbers, determine if the U.S. percentage of the world's population matches the information on the previous page.

3) The *New York Times* article in the Preview Assignment said that in 2008 the U.S. had 751 people behind bars for every 100,000 Americans.

Part A: How is that ratio different from the "One in 110..." described in the excerpt from the ACLU website?

Part B: Use the ratio from the *New York Times* article read in the preview lesson (and the U.S. 2015 population given in question 2) to approximate the number of people behind bars in the U.S. in 2015.

Part C: The *New York Times* article and the ACLU blog said that the U.S. has more than 20% of all prisoners worldwide. Use this information and the approximate number of people in U.S. jails that you just calculated to determine the number incarcerated worldwide.

Part D: What percent of the world's population is incarcerated if we use the world population of 7,276.3 million from 2015?

Part E: Compare the incarceration rate in the U.S. with the incarceration rate worldwide.

---

[2] U.S. Census Bureau. (n.d.). Population Clock. Retrieved October 1, 2015, from http://census.gov/popclock

Lesson 8, Part A More Water, Please!

When preparing for Lesson 3, you read an article about the importance of water to your brain function.[1] The article stated that the "average person in the U.S. drinks less than a quart (32 ounces) of water a day. Yet according to the Mayo Clinic, the average adult loses more than 80 ounces of water every day through sweating, breathing, and eliminating wastes."

Let's say that you are currently drinking 16 ounces of water per day and plan to increase your daily water intake by 4 ounces per week.

Credit:Antonioguillem/Fotolia

1) Determine how many weeks it will take until you reach 80 ounces of daily water consumption. Justify your response.

Objectives for the lesson

You will understand that:

- ☐ Mathematical relationships can be represented in different ways.
- ☐ A linear relationship is characterized by a constant rate of change.

You will be able to:

- ☐ Investigate and compare mathematical relationships using a variety of representations.
- ☐ Create representations to describe mathematical relationships.
- ☐ Write a linear equation given a slope and y-intercept.

The scenario described is an example of a **mathematical relationship** or a relationship between two variables. In this example, one variable is time (measured in weeks) and another variable is daily water intake (measured in ounces).

Mathematical relationships can be represented in four different ways:

- Verbal (as in question 1)

---

[1] Hearn, M. (n.d.). "Water and brain function: How to improve memory and focus." Retrieved August 20, 2014, from http://www.waterbenefitshealth.com/water-and-brain.html.

- Numerical (listing a table of data)
- Graphical (plotting the data and generating a graph)
- Symbolic (algebraic equations)

Each of these modes of describing the relationship is called a **representation**, and the process of using one or more representations to describe a mathematical relationship is called **mathematical modeling**.

2) In our scenario, in Week 0, we are drinking 16 ounces of water per day. As each week progresses, we add 4 ounces to our daily water intake.

Part A: Complete the table of data below for this relationship. Extend the input values for the weeks to complete the table and calculate the daily water intake.

Week	Daily Water Intake
0	16
1	
2	

Part B: What is the amount of change in water intake from week to week?

3) Consider the data points in the table.

Part A: Plot the data from the table on a grid in your notebook.

Part B: Is this linear data? Why or why not?

Part C: Should you connect the points? Why or why not?

Part D: What is the y-intercept of the graph?

Part E: What is the rate of change (or slope) of the graph?

4) Consider the information from the graph.

Part A: Write a linear equation representing the relationship between weeks and ounces of water. Be sure to label your variables!

Part B: Use your equation to calculate the number of weeks it would take to reach an intake of 64 ounces.

5) You have now seen each of the four representations.

The introduction and question 1 is the verbal representation of this scenario.

Question 2 is the ____________________ representation.

Question 3 is the ____________________ representation.

Question 4 is the ____________________ representation.

Use each of the three representations to attempt to answer the question, “How many weeks will it take to reach 80 ounces?”

6) List advantages and disadvantages of each of the four representations.

Lesson 8, Part B What's My Car Worth?

The loss of value over time of something you own, such as a car, is called **depreciation**. This expense is recognized by individuals and businesses for financial reporting and tax purposes. Depreciation expense generally begins when the asset is placed in service.

Credit: Marc Xavier/Fotolia

1) Why do cars lose value over time?

Objectives for the lesson

You will understand that:

- ☐ All proportional relationships are linear, but not all linear relationships are proportional.

You will be able to:

- ☐ Explain the difference between proportional and linear relationships.
- ☐ Explain why the proportionality of changes in two quantities is equivalent to one quantity having a constant rate of change.
- ☐ Compare and contrast linear and proportional relationships.

Recall that two related quantities are **proportional** if the ratio between related pairs of values is constant. For example, in most states, the sales tax and the purchase price of a good are proportional: the ratio of the sales tax to the purchase price is the sales tax rate.

Proportionality is an important component of linear relationships. Some linear relationships are proportional, but some linear relationships are not proportional. If a linear relationship has zero for the y-intercept of the graph, then the quantities and relationship are proportional.

In this lesson, we will illustrate this concept through an example involving the depreciation of a car.

2) Let's say you bought a car for $16,500. The car is expected to last 8 years and then it can be sold for scrap for $500. The value at the end of the life of the car is sometimes called the "salvage value" or "residual value."

Part A: If we assume that the relationship between the value of the car and the year is linear, how much will the value of the car depreciate each year? Explain your strategy.

Part B: Complete the table of values.

Years of Ownership	Value of the Car
0	
1	
2	

Part C: Create a sketch of the data points on your paper. Be sure to label the graph thoroughly. Would it make sense to connect the points?

3) Consider all of the information you have collected in question 2.

Part A: Write an equation for the value of the car in terms of the year.

Part B: What is the slope of the line? Explain what the slope means in this context.

4) Are the value of the car and the number of years since the car was purchased proportional? Justify your response.

5) Look at the changes in the two quantities and the ratio of the change in value to the change in years. Are the changes proportional? How do you know?

6) Examine the proportionality of these two equations:

$$y = 2x + 3$$

$$y = 2x$$

Which is an example of a linear equation that is also proportional? Justify your answer.

Lesson 8, Part C How Money Makes Money

Interest is the amount charged for temporarily using someone else's money. When you borrow money, you must pay back the amount you borrowed (called the **principal**), plus the interest the loan has accrued. To the lender, the interest is income.

In most cases, the interest is a percentage of the amount owed, called the **interest rate**.

1) What is the purpose of interest? Why do banks charge interest on loans and why do they pay interest on savings accounts?

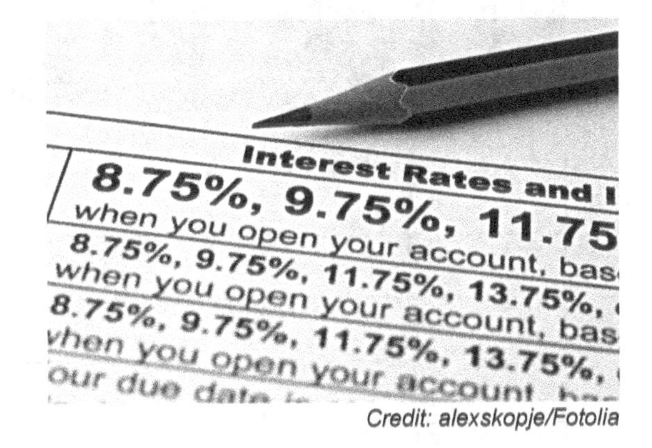

Credit: alexskopje/Fotolia

Objectives for the lesson

You will understand:

- ☐ The difference between simple and compound interest.
- ☐ How interest relates to linear and exponential mathematical models.

You will be able to:

- ☐ Describe the difference between simple and compound interest in practical and mathematical terms.
- ☐ Compare and contrast patterns in linear and exponential models.

There are two types of interest that can be paid to an investor or charged to a borrower: simple interest and compound interest. With simple interest, the amount of the interest is constant throughout the period of the loan or investment. With compound interest, the interest is added to the balance in each investment period, and the percentage of interest is calculated on the new balance in each investment period.

2) Suppose you invest $1,000 in a savings account that pays 2% simple interest annually. When computing simple interest, 2% of the initial balance of $1,000 is added to the account every year. The amount of interest earned is the same each year.

Part A: Complete the following table of the interest and total balance in the account based on the simple interest assumption.

Number of Years (t)	Interest (in dollars)	Total Balance (B) (initial investment + total interest)
Initial deposit		
1		
2		
3		
4		
5		

Part B: Calculate the common difference of the Total Balance between each year.

Part C: Suppose this process continued for 30 years. How much interest will have been earned? How much money would be in the account?

Part D: Write an equation, and then use your equation to determine the amount in the account after 30 years.

Part E: What kind of model is simple interest? Why?

Part F: Sketch a graph of the relationship between total balance and number of years, and then describe the shape of the graph.

3) Now let's suppose that we have the same account, but that interest is **compounded**. We will assume that the compounding occurs annually. Compounding means that each year you will add 2% of the previous year's balance to the account. The amount of interest earned changes each year.

Part A: Use this information to complete the table. Round all dollar amounts to the nearest cent.

Number of Years (t)	Interest (in dollars)	Total Balance (B) (initial investment + total interest)
Initial deposit		
1		
2		
3		
4		
5		

Part B: Suppose this process continued for 30 years. How do we determine how much money would be in the account? Do we have to continue this process for 30 calculations?

Part C: To start the process of finding an equation, find the ratio of each Total Balance amount to the previous Total Balance amount. The first one is started for you:

$$\frac{\textit{Total Balance after 1 year}}{\textit{Total Balance after 0 years}} = \frac{1,020}{1,000} = ?$$

$$\frac{\textit{Total Balance after 2 years}}{\textit{Total Balance after 1 years}} =$$

Part D: Use your answer to Part C and the fact that exponents represent repeated multiplication in order to construct an equation for this relationship.

Part E: Use your equation to determine the amount of money in the account after 30 years.

Part F: This model is called an **exponential model**. Where do you think that name comes from?

Part G: Describe what compounding means from a practical perspective. What does it really mean when we calculate interest on the previous year's balance?

Part H: Sketch a graph of your compound interest model. Describe the shape of the graph.

4) Compare the table from question 2 to the table in question 3.

Part A: Describe the differences that you notice and state which account earns more money for the saver.

Part B: Which model—simple or compound interest—results in a larger balance? What does that suggest about the difference between exponential and linear growth?

Lesson 8, Part D Have My Choices Affected My Learning?

In a previous class, you were asked to rate yourself on the amount you have learned in this class, using a scale from 1 to 10, where:

Learned = 1 means I have learned very little.

Learned = 10 means I have learned almost everything in the course so far.

Credit: Mojzes/Fotolia

1) Do you think that the amount you have learned is related to the amount of time you spend exercising? Sleeping?

Objectives for the lesson

You will understand that:

- ☐ Linear and other models can be used to approximately fit data.
- ☐ Data may suggest a nonlinear relationship.

You will be able to:

- ☐ Use technology to create a scatterplot and estimate the parameters of the line of best fit.
- ☐ Interpret the parameters (slope, y-intercept, correlation of determination) of a simple linear regression.

Let's review and analyze the data that you submitted for the top three categories that affect your learning in this course (from Lesson 1, Part A). The three variables that your class selected are **input** (x) variables and your self-reported rating on Learned is an **output** (y) variable. You will try to figure out how the input variables affect the output variable.

2) Open the Class Spreadsheet to the Student Data tab. Select one of the three variables that you think will have the most significant impact on learning. Highlight that column and highlight the column for Learned. You can highlight two columns that don't touch by highlighting one column, then hold down the Control (or "CTRL") key and highlight the other column.

Generate a scatterplot of your chosen variable and Learned. What do you notice about the scatterplot? Are there any patterns or trends?

Let's determine whether a mathematical model can be used to describe the relationship between these two variables. The process of attempting to fit a mathematical model to data is called **regression**.

It is very rare for a mathematical model to perfectly fit the data. (Remember the slight differences between a perfectly linear model and the Kelley Blue Book car values?) Many factors can have an influence on values of variables. In addition, frequently there is unexplained variation among the values of the variables. This variation is sometimes called **statistical noise**.

The shape of the scatterplot tells you what type of assumptions you can make about the model, which guides what type of model you should use. For example, in early activities for Lesson 8, you saw data that were perfectly linear. You also saw data that were curved and matched an exponential model. You can ask the spreadsheet to show you a **trendline** by using the tabs shown below.

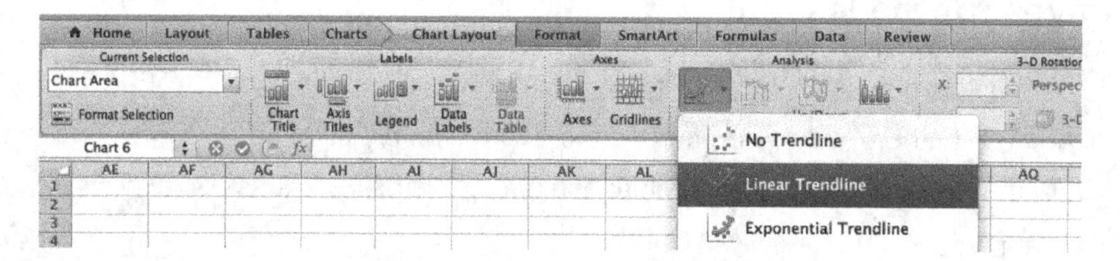

3) Create a linear trendline as shown above. Does the trendline appear to be a good model for the data?

We are going to attempt to use a linear model for our data categories. The spreadsheet regression feature will estimate the slope and intercept of the linear function that fit the data best. This line is called the **line of best fit**. The spreadsheet will also provide some information about how well this line fits the data.

4) In the spreadsheet, use regression to calculate the parameters of the line of best fit. This is called "**running a regression.**" Report the slope and write a practical interpretation below:

Slope:

Interpretation:

The spreadsheet will also tell you values of two statistics that help you determine how well the line fits the data. One statistic is the **correlation coefficient**, denoted by the variable R. The other is the **coefficient of determination**, denoted by R^2 (which is the square of the correlation coefficient).

You will examine these statistics in further detail in later lessons. For now, we will focus on the coefficient of determination, which tells you what percent of the variation in the output variable is due to the input variable.

5) Write the coefficient of determination as a percentage. Then interpret that value by filling in the blanks in the following sentence:

Variable 1: _______% of the variation from the predicted value for Learned is due to changes in the __________ variable.

6) Repeat the steps in questions 3 through 5 for the other two variables. Write an interpretation of the coefficient of determination of each of the two variables by filling in the blanks in the following sentences:

Variable 2: _______% of the variation from the predicted value for Learned is due to changes in the __________ variable.

Variable 3: _______% of the variation from the predicted value for Learned is due to changes in the __________ variable.

7) What conclusions can you draw from your analysis of the three categories/variables and their effects on your learning and performance in this course?

8) How well do you think you have performed in this course? Rate your performance in class on a scale from 1 to 10.

1	2	3	4	5	6	7	8	9	10
My class work is below the instructor's expectations.				My class work usually meets the instructor's expectations.				My class work usually exceeds the instructor's expectations.	

Note that performance is meant to reflect what you have done in class, as opposed to learning. Write your response on your index card and write the word "Performance" next to it.

Lesson 8, Part E Mini-Project: Progressive and Flat Income Tax Systems

An **income tax** is imposed by a government on taxpayers (individuals or organizations) that varies with the income or profits of the taxpayer. Income tax generally is computed as a percentage multiplied by taxable income. The percentage may increase as taxable income increases (referred to as graduated rates). Tax rates may vary by type or characteristics of the taxpayer.

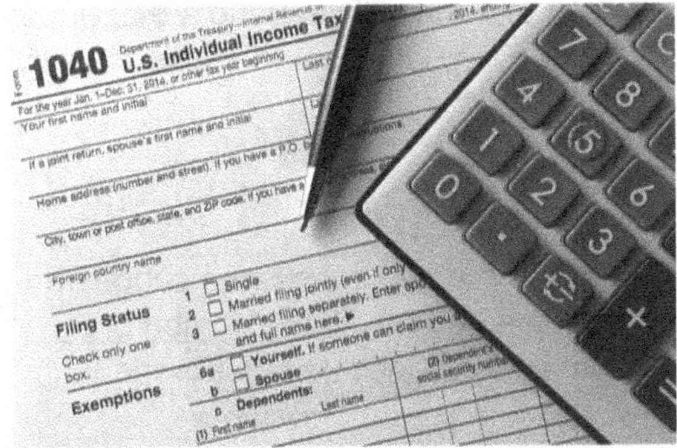

Credit: Kenishirotie/Fotolia

1) Why do governments levy taxes?

Objectives for the lesson

You will understand:

- ☐ The meaning of a progressive income tax system and a flat tax system.
- ☐ That tax policies frequently impact high and low income citizens differently.
- ☐ The basics of current U.S. income tax system and rate structure.

You will be able to:

- ☐ Model a progressive income tax system algebraically and graphically.
- ☐ Compare a progressive income tax system to a flat tax system and identify different outcomes.
- ☐ Explain advantages and disadvantages of different income tax systems.

The United States federal income tax system is **progressive**, which means that higher rates apply to higher incomes.

A **toy example** is a relatively simple problem with an easy-to-understand structure. Let's look at a toy example of a progressive system with simpler numbers that are easy to work with compared with the current U.S. system. (Also, this example does not include deductions and tax credits.) At the end of this mini-project, you will be asked to look up the current rates for the United States.

2) Suppose that annual incomes under $50,000 are taxed at a 10% rate and incomes over $50,000 are taxed at a 20% rate.

   You are currently earning a $49,000 per year salary. You are offered a raise to $51,000 per year. Do you accept the raise? Justify your response, including calculations.

3) In order to avoid the issue suggested by question 2, the 20% tax is not applied to the entire income but only to <u>the portion of the income that is larger than</u> $50,000. This is how progressive income taxes work.

   This is sometimes described by saying that you pay a 10% tax on the first $50,000 of your income and 20% on the rest of your income.

   Reconsider your response to question 2 under this revised system. Calculate your after-tax income before and after the raise and decide whether to accept. Justify your decision, including calculations.

4) Revisit your answers to question 3. If you take the raise, it looks like you would be earning an additional $2,000 but you would actually only receive an additional ________.

5) Using our revised progressive income tax example, calculate the taxes owed on the following incomes:

   Part A: $30,250

   Part B: $45,050

   Part C: $50,000

   Part D: $53,500

Part E: $60,000

Part F: $81,250

6) Look back at your work in question 5. Summarize the steps for the two cases:

- If the gross income is less than $50,000, I . . .
- If the gross income is more than $50,000 I . . .

7) Now convert your work to algebraic form by following the steps given.

Part A: Write an equation for the taxes owed on incomes less than $50,000. Use the variable *G* for Gross Income and *T* for Taxes.

Part B: Your equation should be linear. What is the slope? Write a practical interpretation.

Part C: What is the vertical intercept? Write a practical interpretation.

Part D: Sketch a graph of your equation—only up to $50,000. Let incomes on your horizontal axis extend to $100,000. We will add the second piece of the graph later.

8) The second part of this tax system (the second "bracket") is for incomes over $50,000. In this case, the 20% rate only applies to the amount by which your income exceeds $50,000.

Part A: Write an algebraic expression for the amount by which income exceeds $50,000.

Part B: The income that exceeds $50,000 is taxed at a 20% rate. Use your answer to Part A to write an algebraic expression for the taxes owed on the income that exceeds $50,000.

Part C: For incomes that exceed $50,000, the first $50,000 is taxed at the 10% rate. How much in taxes are owed on the first $50,000?

Part D: The taxes owed on an income above $50,000 are found by adding the taxes on the first $50,000 to the taxes on the amount by which the income exceeds $50,000. Use your answers to Parts B and C above to write an algebraic expression for the taxes owed on an income in this second bracket. Use G for gross income and T for taxes.

Part E: Is the slope of this equation larger or smaller than the slope from question 5?

Part F: Add to your graph from question 5 to illustrate the second tax bracket. Make sure the graph picks up where income is $50,000 and extends to the right. The two "pieces" of the graph should connect when income is $50,000.

9) During a Presidential debate, one of the candidate states, "I propose that we have a 15% flat tax rate."

Part A: What do you think the candidate means by her statement?

Part B: Express the flat tax using a table of values.

Gross Income	Tax Amount

Part C: Express the flat tax with an algebraic equation.

Part D: Create a graph of the equation by putting the line on your original graph in a different color.

Part E: What is so “flat” about the flat tax?

Part F: Broadly speaking, who do you think would pay more in taxes under the flat tax? Who would pay less?

10) The income at the boundary between who pays more and who pays less (the “losers” and “winners”) occurs when the taxes paid under both systems are the same.

Part A: In which bracket should this boundary income be found? Why?

Part B: Set up an equation whose solution is this boundary income. Hint: Set the equation from question 8, Part D equal to the equation from question 9, Part C.

Part C: Solve this equation for G.

Part D: Use your answer to Part C to state precisely which income earners in this example would pay more under the flat tax system and which would pay less.

Part E: What are some advantages and disadvantages of flat tax systems?

11) Using the internet, look up the current rates for single taxpayers in the United States. Report your findings below.

Incomes	Tax Rates
Incomes less than $9,075	10%
Between $9,075 and $36,900	

12) What is the most interesting or useful information you learned in this activity? What are some things you still wonder about the U.S. income tax system?

Lesson 8, Part F Mini-Project: Estimating the Number of People in a Crowd

The City Parks and Recreation Department is sponsoring a musical event to be held at the amphitheater. The event is free but the performer will be paid by the city, based on the estimated number of attendees. Your job is to estimate the size of the crowd.

The area marked with horizontal lines is a grassy space where people often sit. The only information you have about the amphitheater is that the concrete semicircle (shown with vertical lines) has a diameter of 200 feet.

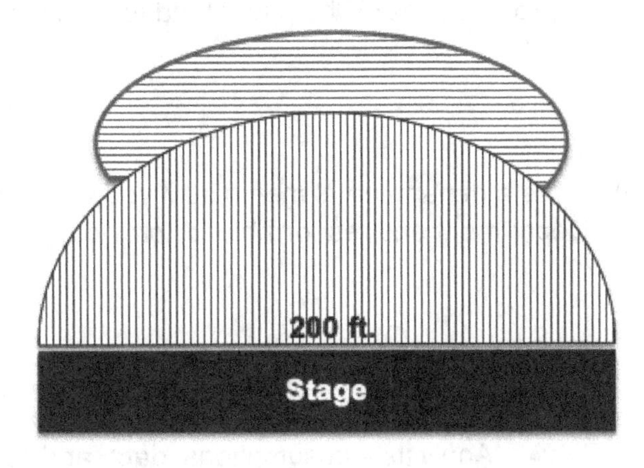

1) What is the crowd usually like near the stage? Farther back? When seated?

Objectives for the lesson

You will understand that:

- ☐ Estimation techniques require reasoning and quantitative analysis.

You will be able to:

- ☐ Use proportions to reason and make estimates.
- ☐ Communicate results with supportive documentation.

There are many reasons why people would like to know the size of a crowd. In this activity, and in the project that follows, you will use geometry and proportions to be able to estimate the size of a crowd.

2) What are your initial thoughts on how to estimate the size of the crowd?

3) Estimate the size of the crowd, using your class approximations.

4) The event was free, but the city promised to pay the performer $0.75 per person, rounded to the nearest $100. Calculate the performer's pay.

5) A lot of guesswork has gone into these estimations. While not insinuating that people would be deliberately dishonest, who in this situation might tend to be generous in their approximations? Who might tend to be conservative in their estimates?

6) Write a departmental memo as if you were an employee of the City Parks and Recreation Department and responsible for providing the estimated attendance and cost.

The memo should have four parts:

- Introduction – what the memo is about and what you are going to say
- Analysis – assumptions, data, and calculations
- Summary – conclusions based on data and why they are meaningful
- Supporting calculations

Example:

Parks and Recreation Department

Official Memo

To:

From:

Re:

Mini-Project

The figure shown is a heat-map of a crowd at a political rally in a town plaza. The darker areas indicate that many people are closely packed in the space. Lighter areas indicate relatively fewer people. The squares are 10' x 10' regions.

You will play three roles in this project:

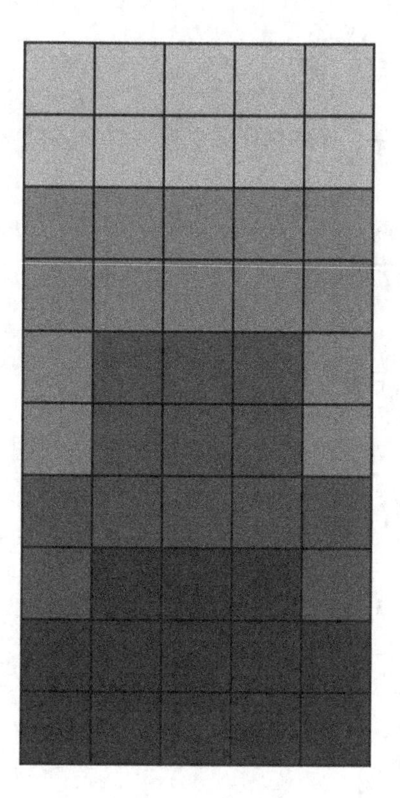

1) You are the media director for the candidate who is speaking at the rally. You are going to provide an estimate of the number of people at the rally for a press release to be sent to the news media. You should use reasonable numbers for the number of people in each region, clearly showing your assumptions and calculations. Then prepare a short press release (3 or 4 sentences).

2) You are the media director for the opposition party. You are going to prepare a press release for the news media, estimating the size of the crowd. You should use reasonable numbers for the number of people in each region, clearly showing your assumptions and calculations. Then prepare a short press release (3 or 4 sentences).

3) You are the political reporter for the local newspaper but you were unable to attend the rally. Using the two press releases you "received," write a news article reporting on the size of the crowd. You should include some numerical support for statements but not include a lot of detailed calculations.

Lesson 9, Part A Depreciation

Depreciation is normally thought of as the decrease in value of an asset. Accountants use depreciation to allocate the cost of a tangible asset over its useful life.

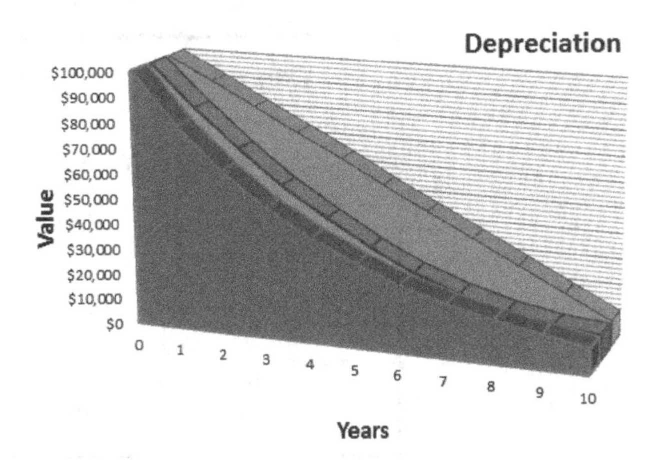

1) Think about the research you did for the Preview Assignment.

Part A: Name some assets that might depreciate.

Part B: Why would a business want to "allocate the cost of a tangible asset over its useful life"?

Part C: Name some assets that might **appreciate**.

Objectives for the lesson

You will understand that:

- ☐ Finding the output value from a graph of a model, given an input value, is an estimate.
- ☐ When given a symbolic model, the exact value of one variable can be obtained given the value of the other variable.

You will be able to:

- ☐ Interpolate and extrapolate using a graphical representation of the relationship between two variables.
- ☐ Use a symbolic model to find the exact value of one variable, given the value of the other variable, and relate those values to the context of the problem.

The Internal Revenue Service defines depreciation as an "annual allowance for the wear and tear, deterioration, or obsolescence of property."[1] Depreciation becomes an income tax deduction that allows taxpayers to recover the cost or other basis of certain property.

---

[1] A brief overview of depreciation. (2015). *Internal Revenue Service.* Retrieved from http://www.irs.gov/Businesses/Small-Businesses-&-Self-Employed/A-Brief-Overview-of-Depreciation

2) A landscaping company purchased a 54" 530cc floating deck lawn mower for $5,450. The graph illustrates the relationship between the years since purchase and the depreciated value of the mower; notice that the three points are **collinear**.

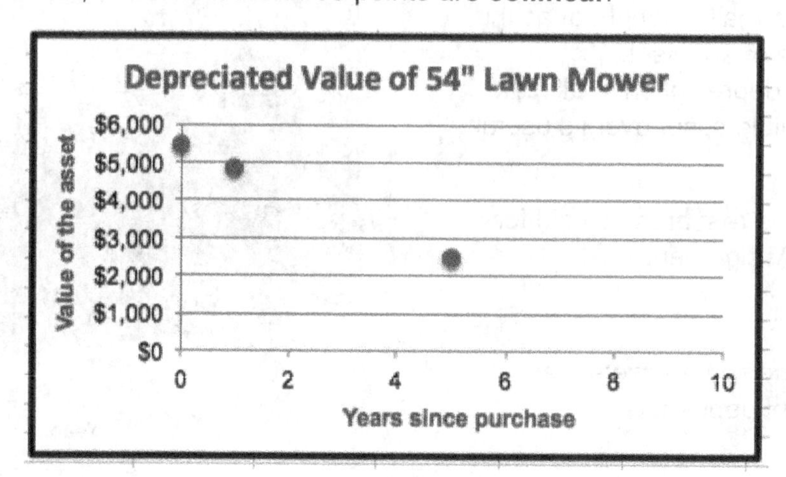

Part A: Estimate the value of the mower at Year 3.

Part B: The company determined that this asset should be depreciated for seven years using the straight-line depreciation method (this means the depreciation is the same amount each year) and a salvage value of $1,250. In Preview Assignment 9.A, you were asked to find the definitions of some terms. Using those definitions and the information given, determine the following information as it applies to the lawn mower.

i) What is the basis of the mower?

ii) What is the class life?

iii) What is the salvage value?

iv) What is the depreciable basis?

v) What is the depreciation for the first year? Round to the nearest dollar, if necessary.

Part C: Complete the table using the graph and depreciation information given.

Years	Value
0	$5,450
1	
5	

Finding a new data point within the range of a discrete set of known data points is called **interpolation**. Since Year 3 is between Year 0 and Year 5, you used interpolation when you found the value of the mower in Year 3.

3) Do you think it would be appropriate to use interpolation to find the value of the mower at Year 2 or Year 4? Explain

4) Use the graph and/or previous information to estimate the value of the mower at Year 6.

Finding a new data point beyond the original observation interval is called **extrapolation**. Since Year 6 is beyond the interval from Year 0 to Year 5, we found the value of the mower in Year 6 by extrapolation.

5) Do you think it would be appropriate to use extrapolation to find the value of the mower at Year 9 or Year 10? Explain. (Watch out—this is a trick question!)

6) An algebraic model can be created to symbolically express the relationship between the age of the mower in years (the input variable) and the value (the output variable) of the mower. The input-output relationship is exhibited in the graphic.

$$Value = 5{,}450 - 600(Years)$$

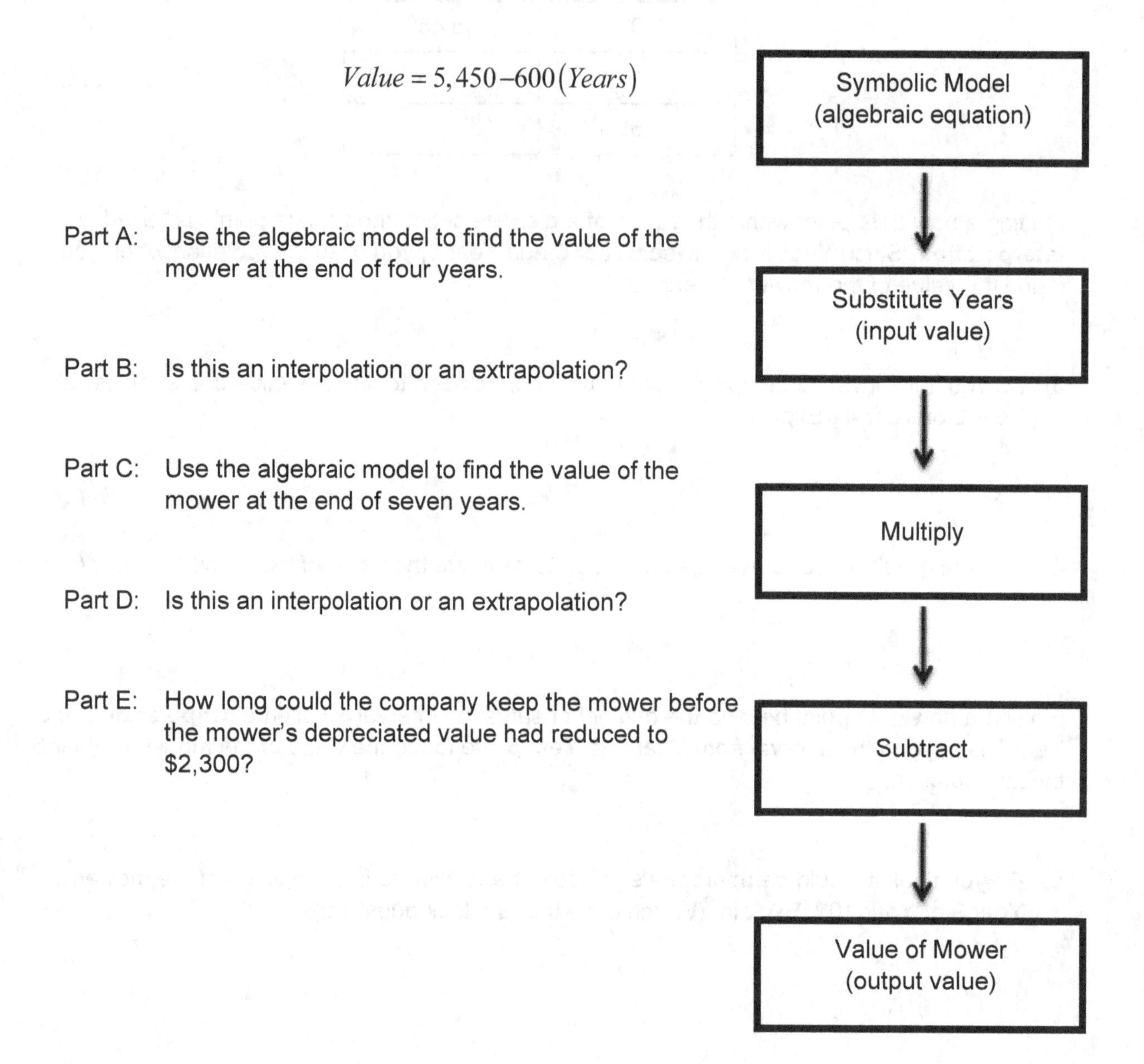

Part A: Use the algebraic model to find the value of the mower at the end of four years.

Part B: Is this an interpolation or an extrapolation?

Part C: Use the algebraic model to find the value of the mower at the end of seven years.

Part D: Is this an interpolation or an extrapolation?

Part E: How long could the company keep the mower before the mower's depreciated value had reduced to $2,300?

7) Without a complete data set, how accurate are predictions about data that you don't have?

Note: In question 6, Part E, you had two options:

1) You could replace $Value$ with \$2,300 and solve for $Years$.

2) You could rearrange the equation to solve for $Years$, and then replace $Value$ and solve.

Regardless which method you use, the end result is that you “undo” the work you did in the previous questions. This relationship is illustrated in the graphic.

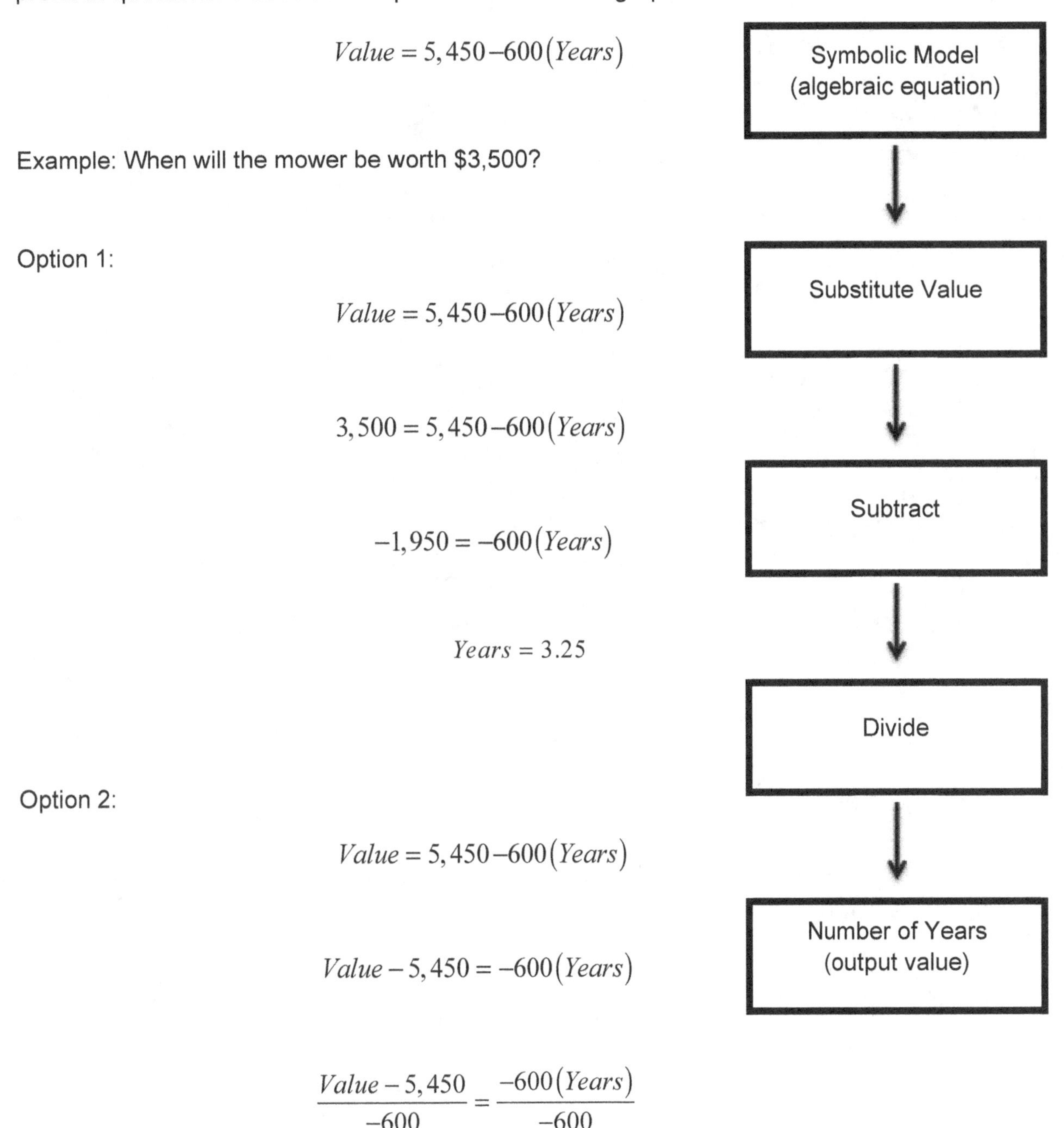

$$Value = 5,450 - 600(Years)$$

Example: When will the mower be worth \$3,500?

Option 1:

$$Value = 5,450 - 600(Years)$$

$$3,500 = 5,450 - 600(Years)$$

$$-1,950 = -600(Years)$$

$$Years = 3.25$$

Option 2:

$$Value = 5,450 - 600(Years)$$

$$Value - 5,450 = -600(Years)$$

$$\frac{Value - 5,450}{-600} = \frac{-600(Years)}{-600}$$

Lesson 9, Part B Appreciating Depreciation

1) In Preview Assignment 9.B, you worked with triangles. How many triangles do you see in the graphic at the right?

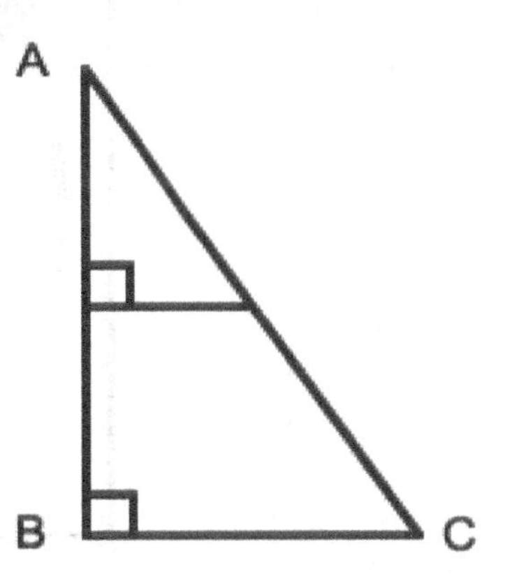

Objectives for the lesson

You will understand:

- ☐ A geometric interpretation of interpolation.

You will be able to:

- ☐ Create a proportion between corresponding sides of similar triangles.
- ☐ Use variables with subscripts.
- ☐ Use the formula for interpolation to find unknown values in a linear relationship.

Two friends sold their vehicles recently. Both vehicles were Dodge Durangos in very good condition and with similar accessories and normal mileage. One was two years old and sold for $26,000 and the other was nine years old and sold for $8,350. In the Preview Assignment, you used this information to estimate a price for a five-year-old Durango.

This lesson will examine the geometry behind the mathematics and introduce you to the formula for linear interpolation using the context of straight-line depreciation.

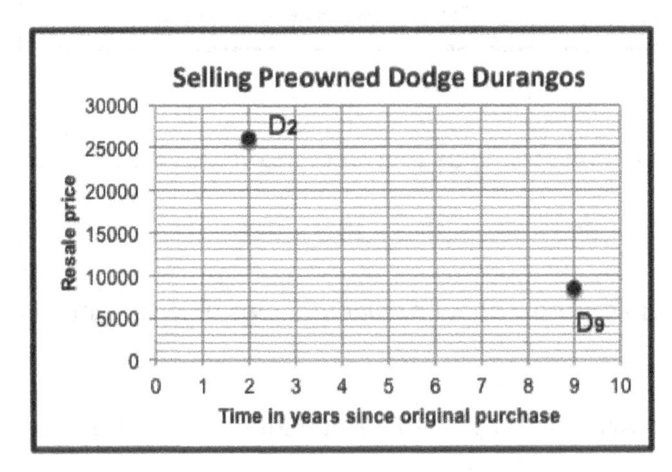

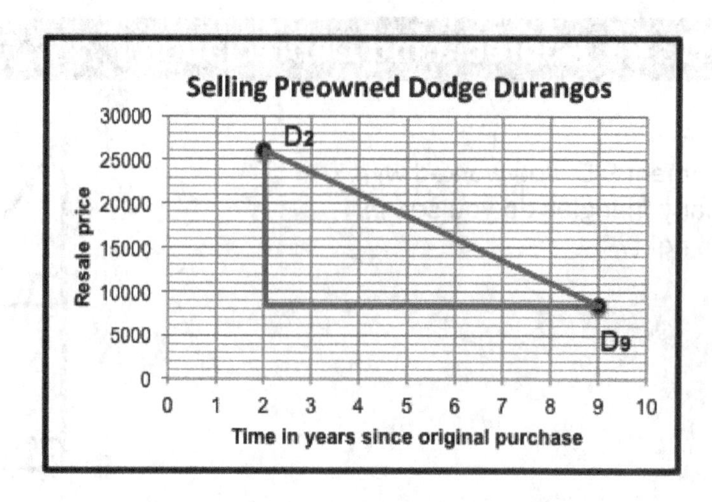

2) What are the height and the base of the triangle shown above? Justify your response.

3) Let's apply the knowledge about similar triangles to the task of estimating prices for preowned Durangos.

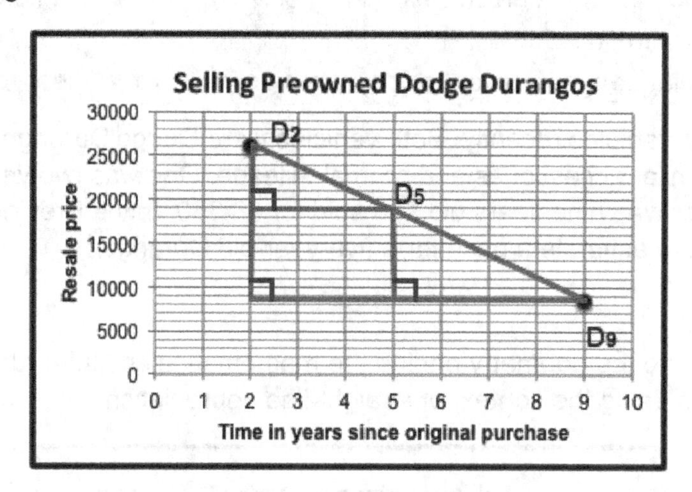

Part A: Draw the separate triangles and label the values that you know with certainty for each one.

Part B: Using what you know about similar triangles, find the height of the top triangle.

Part C: What does this value represent? Put another way, what can it be used for?

4) Recall the coordinate notation from the Preview Assignment, such as $D_2 = (t_2, p_2)$. Use the coordinate notation to represent the work you did in question 3.

Part A: What are the coordinates for D_5 and D_9?

Part B: Using this notation, represent how you found the base lengths of each triangle.

Part C: Using this notation, represent the height of the large triangle.

Part D: Using this notation, represent how you found the height of the top triangle.

Part E: Using this notation, represent how you found the price of D_5.

Your previous work develops the general formula for linear interpolation, using the coordinates of two points, (x_a, y_a) and (x_b, y_b) and the x-value of the intermediate point to find the y-value of the intermediate point.

$$y = y_a - (y_a - y_b)\ \frac{x - x_a}{x_b - x_a}$$

5) Repeat the process of questions 3 and 4, this time using the large triangle and the bottom triangle.

Lesson 9, Part C How Much Should I Be Paid?

When choosing a profession, there are several questions related to finance that could be relevant and need to be considered.

- What starting salary should I expect?
- How much should I expect to be making after five years?
- Will an advanced degree increase my potential salary? If so, by how much? That is, would getting the advanced degree be worth the time, effort, and expense?

1) What other questions would you consider important in anticipating or accepting a job?

8 Tips to Negotiating a Higher Starting Salary[1]

- Do your research.
- Don't tip your hand.
- Understand your value.
- Let the company bring up the salary negotiation issue.
- Emphasize the benefits of your skills.
- Don't blink.
- Be reasonable.
- Be flexible.

Objectives for the lesson

You will understand that:

- ☐ Visual displays, such as scatterplots and histograms, can be used to represent univariate and bivariate data.
- ☐ When two variables are related, the relationship can be described in terms of correlation.

You will be able to:

- ☐ Create a line graph for univariate data.
- ☐ Determine, informally, the correlation between bivariate data.
- ☐ Analyze data and related graphs and describe the trend of the data.

[1] Source: http://www.askmen.com/money/career_100/141_career.html.

Researchers at George Mason University and Temple University conducted a study examining effective negotiating strategies that increased job candidates' starting salary by an average of 5,000: "Assuming an average annual pay increase of 5%, an employee whose starting salary was $55,000 rather than $50,000 would earn an additional $600,000-plus over the course of a 40-year career."[2]

2) One of the assumptions listed in the quote is an annual pay increase of 5%. How likely is it that an employee would average a 5% annual pay increase?

3) A search on the internet can find numerous studies about compensation levels for various professions. Such research is not only valuable to the job seeker, but also to the job provider, since the supply and demand economic model determines, to a large degree, starting salaries.

   The American Society of Civil Engineers and the American Society of Mechanical Engineers conducted a joint engineering salary survey and published their findings.[3] This report includes a great deal of information about salaries of engineers. Read each excerpt and determine the type of data being discussed.

   Part A: "As of March 31, 2012, the average total annual income of respondents in the survey was $103,497 (including salaries, fees, cash bonuses, commissions, and profit received from the respondents' primary jobs during the preceding 12-month period—but specifically excluding overtime pay)." (page 7)

   Is this statistic about univariate or bivariate data?

   Part B: "Full-time salaried respondents holding doctoral degrees in engineering have a median income of $116,000. Those with an M.S. in engineering earn a median of $95,576. Finally, those with a B.S. in engineering earn a median income of $85,900." (page 10)

   Is this statistic about univariate or bivariate data?

---

[2] Marks, M., & Harold, Crystal. (2011). Who asks and who receives in salary negotiation. *Journal of Organizational Behavior, 32*(3), 371–394.

[3] American Society of Civil Engineers and the American Society of Mechanical Engineers. (2012). *The engineering income and salary survey standard report.* Retrieved January 11, 2014, from https://www.asme.org/getmedia/788e990f-99f5-4062-801c-d2ef0586b52d/32673_Engineering_Income_Salary_Survey.aspx. Engineering Income and Salary Survey Publishing Group.

Part C: "Median income shows a consistent increase with increased engineering experience." (page 9)

Is this statistic about univariate or bivariate data?

Part D: "Exhibit 12 reports income by professional responsibility/engineering grade. The graph follows the shape expected, rising from a full-time salaried median income of $55,000 for Engineer I to $150,000 for Engineer VIII." (page 14)

Is this statistic about univariate or bivariate data?

In previous lessons, you examined what statisticians call **univariate data**. That is, you calculated statistics on a single variable—one data set. For example, you studied temperatures in San Francisco and St. Louis and the consumer ratings for treadmills. Univariate data are **quantitative** if the individual observations are numeric values. These variables, in context, usually have units.

Univariate data can be numerically analyzed by finding:

- Measures of central tendency (such as mean or median)
- Measures of spread (such as the range)

We are often interested in how many data points fall within a given attribute, category, or class. Tools include:

- Frequency (and cumulative frequency) tables and graphs
- Dotplots (line plots)
- Histograms
- Box plots

4) The report of the joint engineering survey of salaries mentioned includes the information shown in the table below.

B.S. Degree (Engineering) – Annual Income						
	Number of Responses	10th Percentile	25th Percentile	Median	75th Percentile	90th Percentile
5–9 years of experience	1,317	$56,100	$63,600	$72,300	$84,000	$97,000

Part A: Using the data from the table, construct a number line beginning at $0 and ending at $100,000 with tic marks at every $10,000. Place a solid dot at each of the four percentiles and a solid square at the median.

Part B: Using the same data, construct a new number line beginning at $50,000 and ending at $110,000 with tic marks at every $10,000. Place a solid dot at each of the four percentiles shown and a solid square at the median.

Part C: Compare and contrast the two figures.

Now we will also examine **bivariate data**. That is, a data set that contains two variables.

When examining bivariate data, we often want to know how the two variables relate to each other. For example, educators might want to know if reading scores are related to math scores; that is, whether students who have high reading scores also have high math scores, and vice versa. In an earlier lesson, you read that the statistical term for defining the degree to which two variables are related is called **correlation**. When two variables are strongly linked together, we say they have a high correlation.

Positive correlation: when higher values of one variable are paired with higher values of the other variable. Example: The higher the temperature is outside, the more people show up at the beach.

Negative correlation: when higher values of one variable are paired with lower values of the other variable. Example: The higher the temperature is outside, the less coffee is sold at the coffee shop.

Efforts to describe or summarize bivariate data include creating a scatterplot. The choice of which variable appears on the horizontal and vertical axes is based on which variable is thought to be dependent upon the other. The dependent variable should appear on the vertical axis. When a measure of time (e.g., years of experience) is one of the variables, that variable is most often the independent variable and displayed on the horizontal axis.

5) The report on the joint engineering survey of salaries (questions 3 and 4) also includes the information shown below.

B.S. Degree (Engineering) – Annual Income								
Years of Experience	Under 1 Year	1–2 Years	3–4 Years	5–9 Years	10–14 Years	15–19 Years	20–24 Years	25+ Years
Median Salary in dollars	53,850	57,000	63,000	72,300	89,511	101,255	110,000	122,000

Part A: Define the two variables in this table.

Part B: Which variable is the independent variable?

Part C: Construct a scatterplot of the data.

Part D: Examine your scatterplot. Describe the correlation between the two variables. Would you describe the relationship as positive or negative? Explain.

Part E: What is another way to display these data?

Part F: What other variables might affect this relationship?

For your information

Many websites provide interesting articles about college degrees, starting salaries, and career choice. Selected career information resources that discuss some of these topics are given.

College Degrees and Starting Salaries:

Forbes, "The College Degrees with the Highest Starting Salaries": http://www.forbes.com/sites/susanadams/2013/01/24/college-degrees-with-the-highest-starting-salaries-2/

Campus Explorer, "Degrees with the Highest Starting Salaries": http://www.campusexplorer.com/college-advice-tips/A00B632D/Degrees-With-the-Highest-Starting-Salaries/

Other Considerations:

The Campus Explorer website points out that "while money should factor into your career choice, remember that your passions and interests should be a part of your decision."

Moreover, while a high salary is desirable, there is no guarantee that many of those high-paying jobs are available—or even that a given career is in demand. The Campus Explorer looks at the most in-demand careers in 2014: http://www.campusexplorer.com/college-advice-tips/38E0DE6E/What-Are-the-Most-In-Demand-Careers-in-2014/

Lesson 9, Part D Why Are You Wearing the Same Old Socks?

From the article, "When Science Goes Wrong" that you read in Preview Assignment 9.D, you learned that the proposed solution to SIDS was to shrink the thymus with high doses of radiation or to remove the gland entirely. These procedures had a high mortality rate and led to even more deaths when patients reached young adulthood. These operations were unnecessary and had tragic consequences.

Subsequent research has shown that these doctors were mistaken in their assumptions and that the thymus is not responsible for SIDS. In this case, there was a *correlation* between the size of the thymus and the occurrence of SIDS, but with additional research, it was learned that the enlarged thymus did not *cause* SIDS.

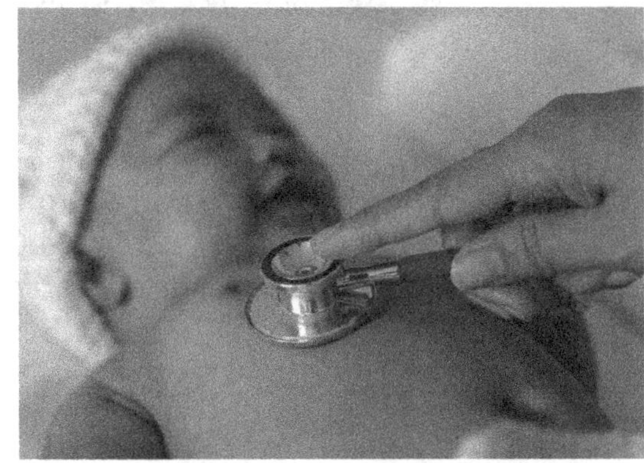

Credit: casanowe/Fotolia

Correlation does not always indicate **causation**.

1) Can you think of two things that seem to be related, where one of them does not *cause* the other one

Objectives for the lesson

You will understand that:

- ☐ Correlation does not imply causation.

You will be able to:

- ☐ Explain why, even if there is a strong correlation, a change in one variable may not cause a change in the other.

Statistical evidence is used to help make decisions about medical procedures, legislation, and educational programs. Just as it is important to learn how to collect and display data, it is also important to learn how to interpret data.

2) Consider the following statements and decide if you think one causes the other, or if perhaps there is a third factor that contributes to the two variables in the statement.

Part A: Someone says, "Every time I eat chocolate, it makes my face break out."

Part B: A newspaper reports, "Recent studies have proven that watching too much violence on TV leads people to be violent in real life."

Part C: A baseball player tells his friends that he did not change socks for three games and the team won all three games. The player concluded that wearing those socks caused the wins and decided not to wear a clean pair of socks for the next game.

Statisticians have defined the **correlation coefficient** as a measure of correlation strength. The correlation coefficient can range from +1.00 (positive correlation) to −1.00 (negative correlation). A correlation coefficient of 0 indicates no correlation.

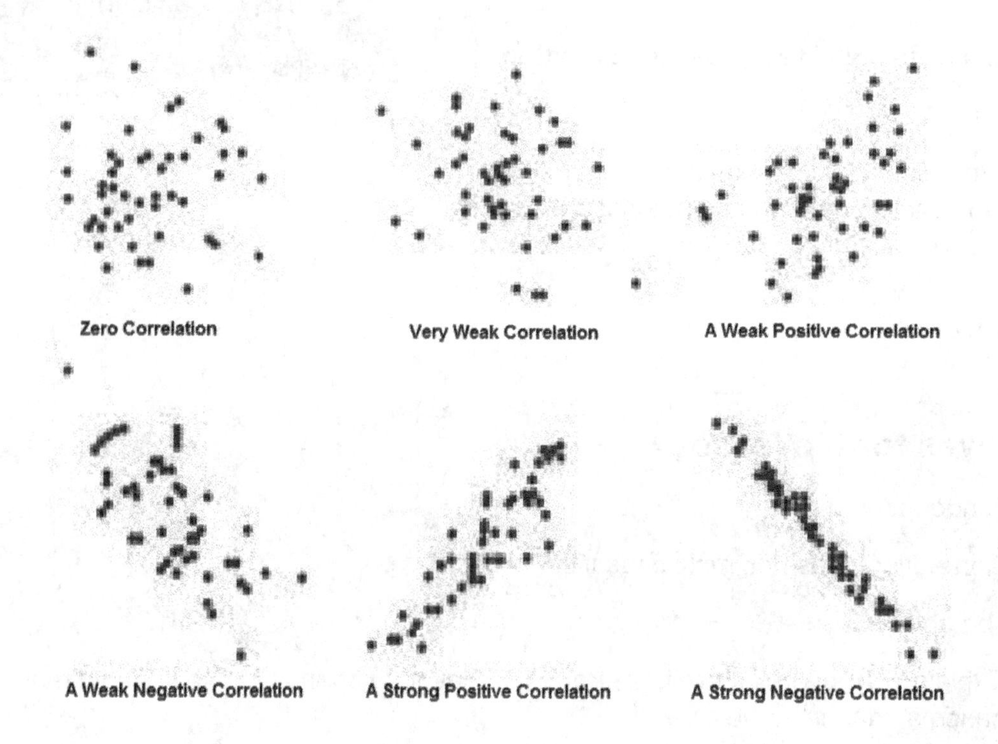

3) Determine whether the correlation is positive, negative, or zero between the variables in each statement below.

Part A: Variable 1: CO_2 levels in the atmosphere since the 1950s.

Variable 2: Obesity levels in the U.S. since 1950.

Part B: Variable 1: The height of school-aged children measured in inches.

Variable 2: The height of school-aged children measured in centimeters.

Part C: Variable 1: The age of a particular model of car.

Variable 2: The value of a particular model of car.

4) A statistics teacher was sitting on a bench overlooking the ocean, eating an ice cream cone. A student in one of his sections sat down next to him and said, "You better be careful—there is a high correlation between eating ice cream and drowning." The teacher was in the middle of a big bite and looked a little confused, so the student elaborated, "Don't you know that days with the most ice cream sales also have the most drownings?"

Does eating ice cream cause drowning? Explain.

5) Data were collected on the weight of cars and their gas mileage. The result is shown in the table below.

Weight (pounds)	MPG (highway)
3489	28
3955	25
3345	27
3085	29
4915	18
4159	21
4289	20
3992	26

Part A: Create a scatterplot of the data.

Part B: Does there appear to be a correlation in the data? If so, is it positive or negative? Weak or strong? Write a contextual sentence about the correlation (if any).

Part C: Name some other factors or "lurking variables" that might also affect the gas mileage of the cars shown.

Part D: Pick any two ordered pairs from the table and find the equation of the line that passes through those two points.

Part E: Use your equation to find the MPG for a car that weighs 2,269 pounds. Is this an example of interpolation or extrapolation? Explain.

Part F: Go to the following Motor Trend website and scroll to the bottom of the page.

http://www.motortrend.com/features/mt_hot_list/1105_11_lightest_2011_model_year_cars_tested/viewall.html

You will find the information for the 2011 Mazda2, which is the vehicle mentioned in Part E. Compare your answer from Part E to the MPG given on the website.

Part G: Interpret the slope and the y-intercept of your equation in context.

Part H: Compare your answers to Part B and Part G.

Lesson 10, Part A Fibonacci's Rabbits

The term *exponential growth* is often used to convey the idea of rapid growth. However, not every instance of rapid growth is exponential growth.

Let's look again at the puzzle posed by Leonardo of Pisa in 1202. Refer to the table of values from Preview Assignment 10.A.

Credit: Grafvision/Fotolia

1) How many rabbits would there be in the thirteenth month if none of the rabbits died?

Objectives for the lesson

You will understand that:

- ☐ Exponential growth is characterized by a rate of growth that is proportional to the population size.

You will be able to:

- ☐ Develop a time series model for the Fibonacci problem.
- ☐ Test whether data are exponential by comparing the rate of growth to the population size.

2) For the moment, assume that the model is an accurate representation of population growth. How many total rabbits (not pairs) do you think there will be at the end of two years? Give an initial "gut-instinct" answer, but then think about ways to get a reasonable estimate that uses some of what you know about how they grew the first year.

3) Use a spreadsheet to determine the number predicted by the model.

The model being constructed in the spreadsheet is a **time series model** as each row in the spreadsheet indicates a single step in time. Time series models can be used in a variety of contexts; for example, budgeting is often done as a time series model where each row indicates a month. We will use time series models of population growth.

Exponential growth is growth that is proportional to the size of the population. Larger populations will grow faster than smaller populations. As a result, not only will the population grow, the rate of growth of the population will grow. This is often referred to as the “snowball effect.” When asked if something is experiencing exponential growth, you should check to see if the rate of growth is proportional to the size of the population.

The (average) rate of change for the rabbit population is equal to the increase in the rabbit population divided by the duration of time that it took to achieve the increase:

$$Average\ rate\ of\ change\ in\ population = \frac{Change\ in\ population}{Change\ in\ time}$$

4) Choose three periods of growth (they may be different months or longer periods but the periods should be the same length) and determine the rate of change of the rabbit population for each of these periods. Are the rates the same? What does this result indicate?

5) Can you find any relationship between the rate of growth of the population and the size of the population? Remember to compare to the “original” population. In other words, if you investigate the change from month 9 to month 10, you should compare to the population in month 9.

6) If the current population is of size 100, can you use this proportionality to find a formula for the population size the following month? What if the population is 1,000? What multiplier should you use to determine the population in the following month from the population in any given month?

7) Can you use this proportionality to estimate the number of rabbits at the end of three years?

8) Assume that you wanted to cap your rabbit population at about 2,000 total rabbits (1,000 pairs) and that it will take two months to implement the procedures to keep the population stable. At what population size should you begin to implement your population control procedures?

Lesson 10, Part B Is It Getting Crowded?

In the last class, we presented a hypothetical example of population growth for rabbits. Now let's use some of those concepts to explore real-world population data.

Credit: *blvdone/Fotolia*

Let's look at the population data and sources that you collected and analyzed in Preview Assignment 10.B. Compare the data and the sources with your group members.

1) Which sources would be the most credible? Which values are most likely to be accurate and which values are most likely to be inaccurate? Why?

Objectives for the lesson

You will understand that:

- ☐ Models are mathematical simplifications of real-world data and phenomena.
- ☐ Different models may be better at matching data than others.
- ☐ More complicated models are introduced to overcome the deficiencies of simpler models.

You will be able to:

- ☐ Evaluate the mathematical appropriateness of a model given historical data.
- ☐ Determine whether a data set suggests a linear or exponential relationship.
- ☐ Use an appropriate model to predict a future outcome.

Now let's examine the world population data from the Preview Assignment 10.B more carefully for signs of linear or exponential growth and then adjust the exponential model to accommodate patterns in the data that are not reflected by the exponential model.

2) Look at the data of the World Population (P) and your table of values from question 1 of Preview Assignment 10.B. Compare the results of your calculations for the first differences and the average rate of growth with others in your group. Determine if a linear or exponential model would fit the data. How do you know?

3) Examine the values of the average rate of growth. If the data represent an exponential relationship, what characteristic do you expect to see in the values of this column? Why do the actual values differ from the expected values?

4) Assume that $\frac{R}{P} = k \times (L - P)$ for some unknown values of k and L. Solve the equation for R. (Note: There will be more discussion about k and L in later lessons.)

5) Complete the table of values and sketch a graph of your formula in question 4, assuming that k = 0.1 and L = 100.

P	
0	
25	
50	
75	
100	

6) Assuming that the formula used in questions 4 and 5 is correct, answer the following questions.

Part A: What can you say about population growth when the population is very close to 0? What would account for this?

Part B: What can you say about population growth when the population is very close to L? What would account for this?

Part C: What can you say about population growth when the population is very close to the middle of the graph? What would account for this?

7) The graph you completed in question 5 relates the rate of growth to the population size. We are interested in determining how the population changes over time.

Part A: Start with a small population at a given time. How will the population change? How will this cause the rate of the population growth to change?

Part B: Now assume that the population has grown significantly, so that it is getting close to size L. How will the population change? How will this cause the rate of the population growth to change?

Part C: What is it about the equation that causes this behavior in the growth rate?

Part D: Although time increments are not included in the table in question 5, try to create a rough sketch of the population size over time.

Lesson 11, Part A Population Growth

As you studied exponential growth, you probably said more than once that things can't just continue to grow exponentially. Whether it is a population, energy consumption, the rate of deforestation, or even internet use, most things that grow at an exponential rate can't continue to do so.

Consider a metropolitan area that has been growing at a relatively constant rate of three to four percent each year.

Credit: Andrea Izzotti/Fotolia

1) When roads get too crowded, water and sewer capabilities are maxed, and the population gets so dense that safety and health are compromised, what will happen to the population growth?

Objectives for the lesson

You will understand that:

- ☐ A population that grows at an increasing rate and then continues to grow at a decreasing rate might be modeled logistically.

You will be able to:

- ☐ Sketch a model for a population that increases at an increasing rate.
- ☐ Sketch a model for a population that increases at a decreasing rate.
- ☐ Identify behavior in a graph, draw conclusions about the behavior, and predict future outcomes.

Smith Island is a complex of three parallel semi-tropical islands located in Brunswick County, North Carolina. This county has been a popular site for television series such as *Dawson's Creek,* and movies such as *Weekend at Bernie's* and *I Know What You Did Last Summer.*

The deer herd on Smith Island is a recent population, originating sometime in the mid-1980s. The herd was estimated at 30 in 1996. Hunting is not allowed and natural predators are practically nonexistent, except for alligators that occupy freshwater ponds. Thus, the population

of deer increased at an increasing rate in the years following 1996.[1]
Suppose that in 2002, the study found that the density of deer was approximately 19.5 deer per square kilometer. (The area of Smith Island is about 48 square kilometers.) Estimates, based on research, showed that the densities ranged from 11.9 deer/km^2 in some areas to 57.8 deer/km^2 in other areas.[2]

2) According to the 2002 study, what was the deer population on Smith Island in that year?

3) Let's create some representations of changes in the deer population over time.

Part A: Using the information from the two paragraphs above about the deer population on Smith Island, complete the following table.

Year	Number of Deer
1980	
1996	
2002	

Part B: Make a quick hand sketch of the data in the table. Should these points be connected? Why or why not?

Part C: Describe (in general) the rate of change of the deer population.

---

[1] Source: http://www.bhic.org/wildlife.

[2] Ray, D. K., Golen, E. G., & Webster, W. D. (2001). Characteristics of a barrier island deer population in the southeastern United States. *The Journal of the Elisha Mitchell Scientific Society, 117*(2), 113–122. Retrieved October 15, 2014, from http://dc.lib.unc.edu/cgi-bin/showfile.exe?CISOROOT=/jncas&CISOPTR=3656.

The islands, especially Bald Head Island, continued to develop and became more urban, with developments replacing farms and fields. Residents began to worry that the deer population would destroy the maritime forest as well as ornamental plants found naturally. There were also an increasing number of deer-automobile conflicts.

4) Describe what you think would happen to the deer population on Bald Head Island between 2002 and 2010 without any intervention.

5) A **parameter** is a value that helps define a relationship. One example of a parameter is the carrying capacity of a species. The **carrying capacity** in a particular environment is the maximum number of that species that the environment can support, due to food, water, and other environmental limitations. The carrying capacity of deer on Smith Island has been estimated to be about 33 deer/km^2.

Part A: Let's say the deer are projected to reach the carrying capacity by 2020. What will be the population of deer on Smith Island? Add your answer to the table in question 3, Part A.

Part B: Describe what you think happens to the rate of increase in deer during the years leading up to the point when the deer population reaches its carrying capacity.

Part C: Sketch a curve that you believe best represents the deer population on Smith Island from 1980 through 2020.

Part D: Compare the curve you drew with those drawn by your group members. What things are essentially the same? What things are a little different?

Culling is the process of removing breeding animals from a group based on specific criteria. This may be done to reinforce certain desirable characteristics or to simply reduce the number of animals.

The Village Council approved the culling of deer on the island. The goal was to preserve the health of the herd and to protect island ecology from damages of over-grazing. Culls have taken place several times since 2001.

6) Knowing now that actions were taken to decrease the population of deer, sketch a new curve for the deer population from 1980 through 2020.

Lesson 11, Part B Oh Deer!

In this lesson, we will derive the symbolic (algebraic) form of the deer population logistic model. The concepts and explorations will provide insight into how mathematicians investigate numerical data, such as carrying capacity and logistic growth rate, to find a symbolic representation of the model.

During this activity, you will need to have your work from Lesson 11, Part A, for reference.

1) Look back at your sketch from Lesson 11, Part A, question 5, Part C. When was the growth rate at its lowest? When was the growth rate at its highest?

Credit: Steve Oehlenschlager/Fotolia

Objectives for the lesson

You will understand that:

- ☐ Logistic curves and symbolic models can be used to approximate some real-world data.
- ☐ A logistic model uses historical data to determine the parameters **carrying capacity** and **logistic growth rate**.

You will be able to:

- ☐ Develop discrete models of natural phenomena and use the models to predict future values.
- ☐ Calculate the carrying capacity and logistic growth rate of a real-world scenario.

2) In the Lesson 11, Part A activity, you analyzed the data to create a graphical representation. In this activity, you will use the data to create a symbolic representation. You will determine the parameters of a logistic model that best matches the data.

Part A: What was the estimated carrying capacity for the deer on Smith Island? Call this amount L.

Part B: Recall that in 1996, there were 30 deer. This number represents what percentage of the total carrying capacity? Round to the nearest whole percentage point.

Part C: In 1996, what percentage of the carrying capacity remained?

Part D: Let P_t = the population of deer at time t. Show your calculations from Part C using the variables L and P_t instead of numbers.

3) As you saw in Preview Assignment 11.B, when studying two discrete points in time, relative change in population can be expressed as $\frac{\Delta P}{P}$. Recall that this is called the **growth rate**.

In a logistic model, the growth rate at time t is proportional to the remaining capacity. The constant of proportionality is designated by the letter k.

$$\frac{\Delta P}{P} = k \times \left(1 - \frac{P}{L}\right)$$

The growth amount (absolute change in growth) is called the **incremental growth** and can be expressed as:

$$\Delta P = k \times P \times \left(1 - \frac{P}{L}\right)$$

Part A: How is the second formula developed from the first formula?

Part B: Let's look at two special situations, when $P = 0$ and when P is near L. Looking at the right-hand side of the formula for $\frac{\Delta P}{P} = k \times \left(1 - \frac{P}{L}\right)$, what will be the value of $\frac{\Delta P}{P}$ when $P = 0$?

$\frac{\Delta P}{P} =$

Part C: Looking at the right-hand side of the formula for $\frac{\Delta P}{P} = k \times \left(1 - \frac{P}{L}\right)$, what will be the value of $\frac{\Delta P}{P}$ when the value of P approaches the value of L? In your own words, explain what is happening in this situation.

$\frac{\Delta P}{P} =$

Part D: In the real world, it is possible for the population to exceed the carrying capacity; models provide projections and are usually not precise. When $P > L$, what does that tell you about $\frac{P}{L}$ and about $\left(1 - \frac{P}{L}\right)$? What would it mean with regard to ΔP?

Lesson 11, Part C Can You Hear Me Now?

In 1959, Arthur C. Clarke predicted a "personal transceiver, so small and compact that every man carries one . . . the time will come when we will be able to call a person anywhere on Earth merely by dialing a number."[1]

It took over 35 years from when the concept of cellular phones emerged, in 1947, for the technology to be commercially available in the United States. By 1987, cellular telephone subscribers exceeded one million and the airwaves were crowded.[2] According to a Pew Research study in 2013, over 90% of American adults owned at least one cell phone.[3]

Credit: *william87/Fotolia*

1) Make a quick sketch of a graph that you believe shows the growth of cell phone usage in the U.S.

Objectives for the lesson

You will understand that:

- ☐ A change in a parameter of a logistic model changes the shape of the graph and the output value.
- ☐ Initial population, growth rate, and carrying capacity are the parameters of the logistic model in this lesson.

You will be able to:

- ☐ Explore the changes in the values of the parameters of a logistic growth model and describe the effect of those changes on the model.

The curve that best models the growth of the deer population on Smith Island is also the curve that best models many populations that grow to some limit. This curve is used in biology,

---

[1] Clarke, A. (1962). *Profiles of the future: An inquiry into the limits of the possible.* New York: Harper & Row.

[2] Source: http://inventors.about.com/library/weekly/aa070899.htm.

[3] Rainie, L. (2013, June 6). Cell ownership hits 91% of adults. Retrieved October 14, 2014, from http://www.pewresearch.org/fact-tank/2013/06/06/cell-phone-ownership-hits-91-of-adults.

chemistry, and in many other fields; it is also useful in advanced statistics. From the Preview Assignment 11.C, you know that the growth in cell phone subscriptions is slowing down.

In a previous lesson, you were working with a **recursive model**, in which you had to know the previous population value in order to predict the next population value. Wouldn't it be more convenient to be able to jump ahead several steps into the future without calculating all the steps in between?

An **explicit equation** allows you to do just that: you can substitute in any input value and calculate the corresponding output value. You have used this method many times with linear equations.

The explicit form of the logistic model is shown in the spreadsheet image below. $A, B,$ and C are parameters for the equation. You will graphically investigate the effect of changing these parameters.

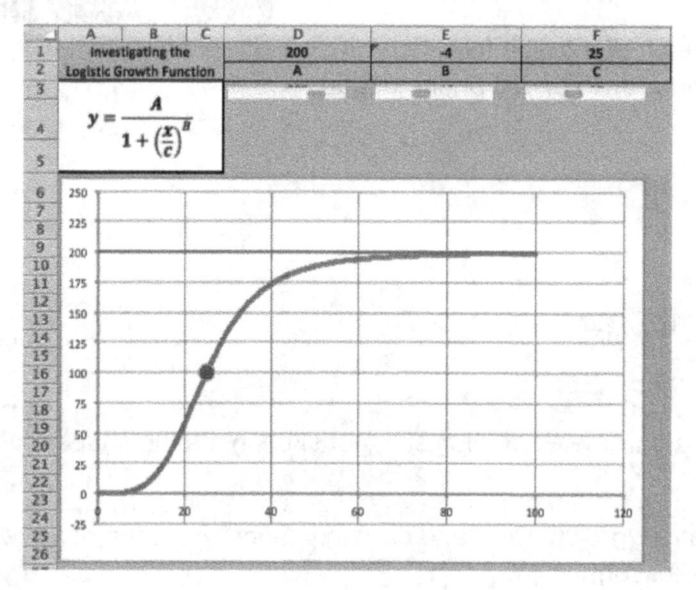

2) Open Spreadsheet_11C_Investigating_the_Logistic_Function. Take a moment to familiarize yourself with the graph and sliders. Make sure the values are set at $A = 200$, $B = -4$, and $C = 25$.

Part A: Label the horizontal line at $y = 200$ on the image above as an "**asymptote**." In this case, the line represents the carrying capacity or maximum value of 200. (Notice that the curve approaches the horizontal line but does not seem to cross it, which is the basic idea of an asymptote).

Part B: Use the slider to change the value of A (the value shown in Cell D1). Watch the changes to the curve as you do so. How does the value of A affect the shape of the curve? Write your conclusions in complete sentences. For example, one sentence might begin with the phrase "As the value of A decreases from 200." Another sentence might start with the phrase "As the value of A increases from 200." A third sentence might summarize your findings.

Part C: Put the value of A back to 200. Now use the slider that controls the value of B, watching the curve as you do so. How does the value of B affect the shape of the curve?

Part D: Return the value of B back to –4. Use the slider that controls the value of C, watching the curve as you do so. How does the value of C affect the shape of the curve?

Part E: What significant change in the shape of the curve seems to happen at the large point? That is, how are the two portions of the curve (to the left of the point and to the right of the point) different?

Part F: As you investigated each of the three parameters, describe what happened to the large point.

3) The number of deer on a fixed space (an island) will likely have an upper limit. A logistic model is appropriate. The population will grow slowly as it reaches that upper limit (carrying capacity). The logistic model does not work as well for the number of cell phones in use in the United States. In what ways is the logistic model appropriate for the deer population and in what ways is the model less appropriate for the number of cell phones?

Lesson 11, Part D Hares and Lynxes

In the logistic models in previous lessons, we assumed a fixed carrying capacity for a population. However, other environmental factors may affect populations and prove to be more important.

Credit: impr2003/Fotolia

1) Consider the hares and the lynxes in the article. Why did the logistic and exponential models fail?

Objectives for the lesson

You will understand that:

- ☐ Complex models, such as predator/prey models, may be neither logistic nor exponential.
- ☐ The mathematical concepts of proportionality can be combined with other mathematics concepts to model complex behavior.

You will be able to:

- ☐ Identify the constant of proportionality in a real-world scenario.
- ☐ Develop a parameterized time series model with more than two dependent variables in a spreadsheet.

Using the information from the article you read in the Preview Assignment ("What Drives the 10-year Cycle of Snowshoe Hares?"), we will develop a two-variable time series model to predict population size based on the current population size for two species that interact.

2) Answer the following questions based on the article.

   Part A: Is it reasonable to assume that the hare population never gets large enough to lead to resource constraints?

   Part B: Let's begin by investigating a simplified model for hare population growth, assuming that the population never strains the available resources and there are no predator lynxes. If there were 100 hares in the initial population and 102 hares at the end of the first month, what is b, the hare population growth rate for the month?

Part C: Write a general formula that determines ΔH, the number of hares added to the population in one month if there were H hares at the start of the month and the growth rate stays the same as Part B?

3) Now let's focus on the population of lynxes. In the absence of food (hares), the lynx population will experience **exponential decay**.

Part A: If there were 100 lynxes in the initial population and 99 lynxes at the end of the first month, what is m, the monthly mortality rate?

Part B: Write a general formula that determines ΔL, the number of lynxes subtracted from the population in one month if there were L lynxes at the start of the month and the mortality rate stays the same as Part B.

4) So far, we have a pair of exponential models. As they stand, how accurate would they be for long-term predictions if the hares and lynxes did not interact with each other?

5) Let's see what happens when the hares and the lynxes interact. In particular, whenever they encounter each other, there will be a slight chance that the lynx will kill the hare.

Part A: If the number of hares increases, how will this affect the number of lynx-hare interactions? If the number of lynxes increases, how will this affect the number of lynx-hare interactions?

Part B: Assume that the number of kills (a lynx killing a hare) is jointly proportional to both the number of hares and to the number of lynxes. Using k as the constant of proportionality, derive a formula for N, the number of kills in a month with H hares and L lynxes.

Part C: If there are 100 hares and 10 lynxes, there will be 2 kills in a month. Using the formula you developed in Part B, determine the value of k.

Part D: Using your formula from Part B and the value of k found in Part C, write a general formula for N, the number of kills in a month with H hares and L lynxes.

6) Now let's derive a formula for the change in the hare population when the two populations are interacting. At the start of the month, you have H hares and L lynxes. The hares reproduce as in question 2, but are killed by lynxes as described in question 5. What is the change in the hare population (ΔH) in that month?

7) Let's look at the change in the lynx population when the two populations are interacting.

Part A: Assume that the lynx population requires 20 kills to allow it to grow in size by one lynx. What is G, the new growth in the lynx population?

Part B: If at the start of the month you have H hares and L lynxes, and the lynxes die out as in question 3 but are also growing by killing the hares as described in question 7, Part A, what is the change in the lynx population (L) in that month?

The equations developed above will be used in the spreadsheet template to be completed in Practice Assignment 11.D.

Lesson 11, Part E Reindeer and Lichens

Population growth and decay can be influenced by many factors. Considering these factors in our models helps us to develop a model that is as accurate as possible.

In Preview Assignment 11.E, you began an exploration of the fate of the reindeer on St. Matthews Island, Alaska.

1) How is the relationship between reindeer and lichens similar to or different from the relationship between hares and lynxes?

Credit: dinozzaver/Fotolia

Objectives for the lesson

You will understand:

- ☐ How to analyze and determine the appropriate model for a real-life scenario.

You will be able to:

- ☐ Determine parameters to match a model's predictions against historical data.
- ☐ Create a spreadsheet involving the formulas of the model to predict future behavior.
- ☐ Adjust models based on rounding to account for rounding.

Our work in the last lesson resulted in a cyclic model of population growth as two species interacted. In this lesson, we have another model that is driven by two species, but it appears to exhibit quite different behavior. Our task is to develop a predator-prey model that will represent the lichens and reindeer on St. Matthew Island. This model will have to be modified from the animal models that we have seen, since the population size of lichens must be measured in a different way.

2) If we are trying to match the lichens and reindeer to the predator-prey model in Lesson 11, Part D, which species plays the role of the prey? Which species plays the role of the predator?

3) Think about the parameters of the hare-lynx problem. Identifying the parameters that are similar in the two situations will help to develop the new model. Complete the following:

Part A: The birth rate of the hares is similar to _______________.

Part B: The mortality rate of the lynxes is similar to _______________.

Part C: The kill rate of the hares by the lynxes is similar to _______________.

4) On St. Matthew Island, the reindeer consumed all of the lichens and virtually died out. Think about the parameters in the predator-prey model. How do the environmental conditions for the reindeer and lichens relate to or impact those parameters?

5) Open Spreadsheet_11.E. Notice that it contains the original hare-lynx information from the previous activity. You will adapt the spreadsheet to create the reindeer-lichen model.

Let's assume that our model includes an initial reindeer population (R) of 30, amount of lichens (L) is 200, lichen growth rate (b) is 0.02, reindeer mortality rate (m) is 0.01, grazing rate (k) is 0.004, and the fertility rate is 20. Adjust the spreadsheet to reflect the new names and values. Take care that you replace the hare and lynx information according to your answers to questions 2 and 3 above.

Part A: According to the spreadsheet, what happens to the lichen and reindeer populations?

Part B: Create a line graph containing the reindeer and lichen data. Does the graph match your answer to Part A? What do you think of these graphs of your algebraic representations and their relationship to what really happened with the reindeer and lichens?

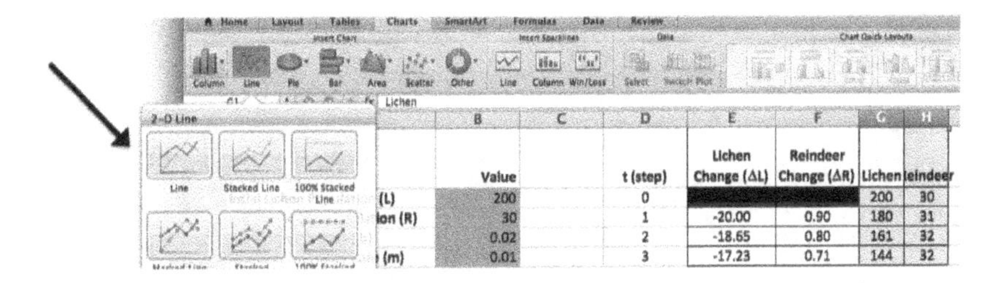

6) What changes could you make to the parameters to keep disaster from happening?

Lesson 12, Part A How Long Is the Longest Day?

A **periodic function** is a relationship with a repeated pattern. Graphs of periodic functions can have many shapes, as long as the graph repeats itself. You saw these cyclical patterns in ocean tides in Preview Assignment 12.A and in the predator-prey populations in Lesson 11.

1) Locate the graph of the tides that you created in the Preview Assignment. Discuss how the behavior of the graph relates to the tides in Galveston, Texas.

Credit: ArtFamily/Fotolia

Objectives for the lesson

You will understand;

- ☐ What a periodic model is and the vocabulary that describes this model.

You will be able to:

- ☐ Sketch a graph that depicts a periodic phenomenon.
- ☐ Identify the period and amplitude of a periodic function.
- ☐ Compare and contrast the graphs of different periodic models.

Applications of periodic behavior can be found in fluid flow, wave motion (e.g., sound and current), economics, and business (e.g., crop yields and prices).

In a periodic function, the **period** (P) is the length of time that it takes for the cycle to repeat itself (example shown at the right).

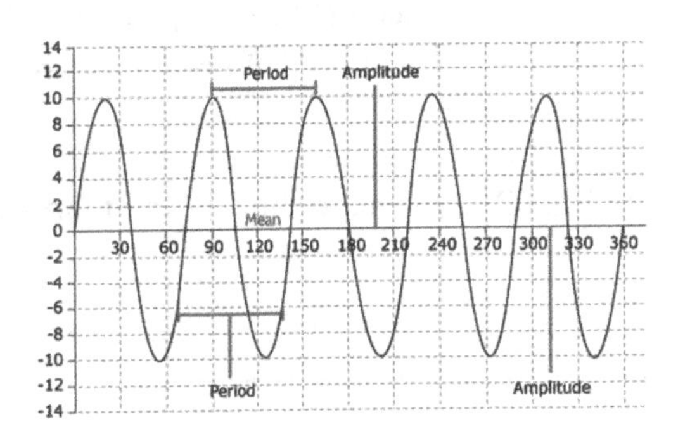

The **amplitude** of a graph is:

$$A = \frac{1}{2}(\text{maximum height} - \text{minimum height})$$

2) Consider the graph of the high and low tide in the Preview Assignment.

Part A: What is the period of the graph?

Part B: What is the amplitude?

3) The number of daylight hours in a particular location depends upon latitude. The variation is caused by the tilt of the Earth's axis of rotation with respect to the ecliptic plane of the Earth around the Sun. The graph below shows a prediction of the number of daylight hours for 2016 and part of 2017 in New Orleans, Louisiana.[1].

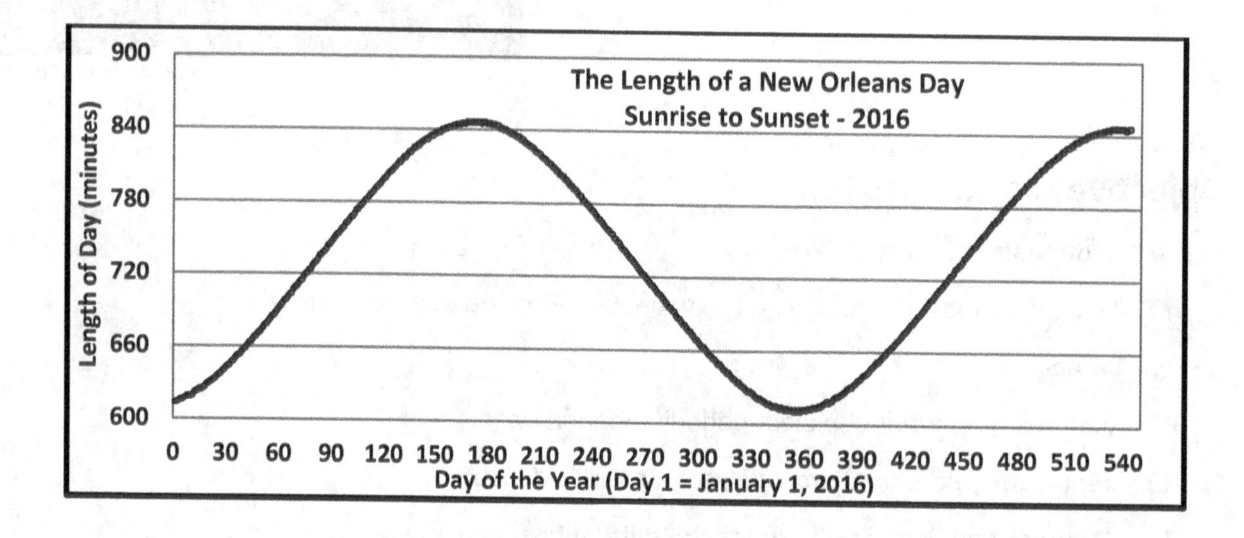

Part A: How many minutes pass between two consecutive sunrises?

Part B: Estimate the length of the shortest day (sunrise to sunset) in New Orleans in 2016. On approximately what day will it occur?

Part C: Estimate the length of the longest day. On approximately what day will it occur?

---

[1] The predictions for the time of sunrise and sunset are fairly reliable. See http://www.sunearthtools.com/solar/sunrise-sunset-calendar.php.

Part D: What is the length of the average day in New Orleans? Draw a horizontal line through the graph to represent the average length of a day.

Part E: There are at least two ways to determine the amplitude of this periodic function. Choose one way, and find the amplitude.

Part F: The curve exhibits a repeating pattern. Place a point on the curve at Day 80. Place a second point on the curve when the length of the day is the same as it was on Day 80 and the length of the day is changing in the same way that the length of the day is changing on Day 80. How many days are there between these two points?

Part G: The length of time that it takes the curve to repeat is called the period of the curve. What is the contextual significance of the period of this curve?

Part H: The number of daylight hours changes more rapidly during parts of the year and more slowly during other parts of the year. When is the curve changing rapidly? When is it changing slowly?

4) The graph showing the length of a New Orleans day is repeated below.

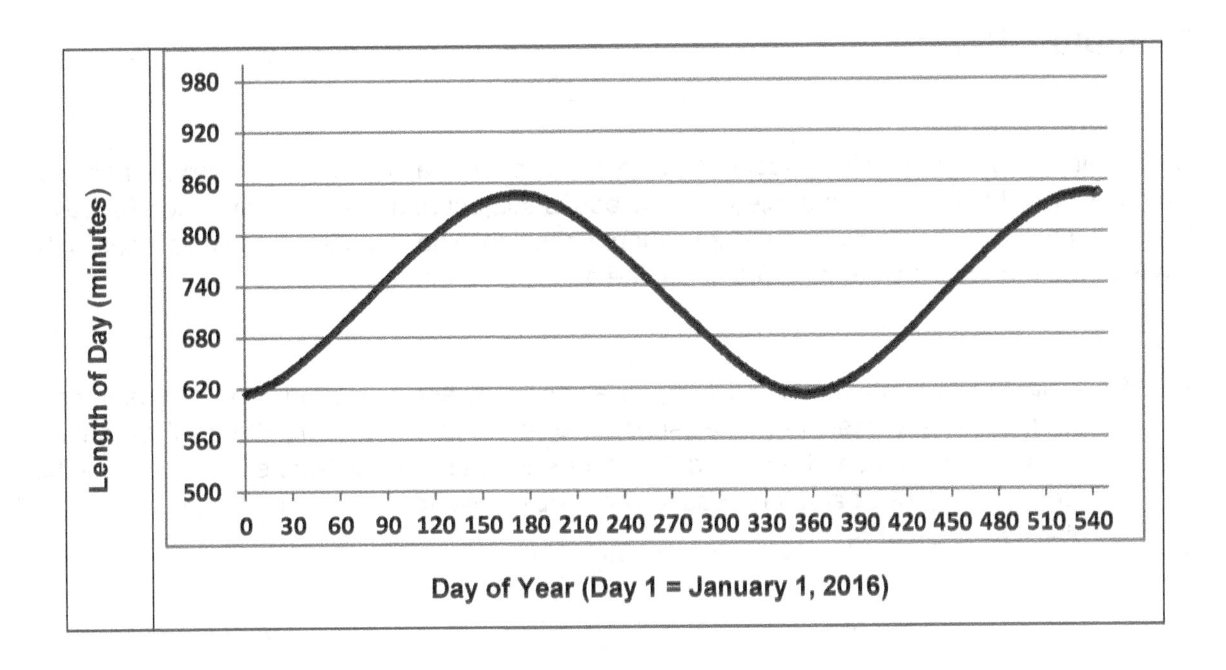

Part A: Do some internet research to determine the lengths of the longest and shortest days of 2016 in Seattle, Washington, and when they occur. Sketch a graph of the length of a day in Seattle on the same axes as the New Orleans graph. Use a different color or a dashed line.

Part B: What is the period of the Seattle graph? What is the amplitude?

Part C: Research information about the lengths of the longest and shortest days of 2016 in Santiago, Chile, and when they occur. Sketch a graph of the length of each day in Santiago on the same axes as the New Orleans and Seattle graph. Use a different color or a dashed line.

Part D: Compare and contrast the New Orleans graph with the Santiago graph.

If you are interested in more information about day length, visit:

- http://www.smithsonianmag.com/smart-news/weird-blips-randomly-change-the-length-of-earths-days-for-months-on-end-10661780/
- vimeo.com/24756892

Pendulum Behavior

A pendulum is a bob (a weight) suspended from a pivot so that it can swing freely. When the weight is pulled to one side and released, the bob is subject to a restoring force due to gravity that will accelerate it back toward the equilibrium position. The restoring force combined with the mass of the bob causes it to oscillate about the equilibrium point.

Part A: Construct a pendulum using string or a shoelace, and something you can tie to one end, such as a washer or small superhero action figure. (A long string will be easier to work with than a short one.) Tie the other end to something stable that will allow the bob to swing back and forth easily. If possible, use your cellphone to take a picture of the pendulum at rest.

Part B: Pull the bob to a certain point and let go. If possible, take a picture of the pendulum in action.

Notice how long it takes to make a complete cycle and about how far the pendulum travels from one side of equilibrium to the other. Repeat the experiment several times, pulling the bob to about the same location you did initially. Repetition will help you make better estimates of the period and amplitude of the pendulum.

Part C: Record the time and distance in the table of values. You need at least five data points: the point when the bob is releases ($t = 0$); the approximate time when the bob moves through the equilibrium point the first time (when $d = 0$); the approximate time and distance the bob reaches a maximum distance from equilibrium on the other side; the approximate time when the bob moves through equilibrium the second time (when $d = 0$ again); and the approximate time when the bob returns to its original position. Record the distance from center to the right of center as positive values and to the left of center as negative values.

Time in Seconds	Distance from Center in Inches
0	
	0
	0

Part D: Create a graph to show the movement of the pendulum bob.

Lesson 12, Part B What's My Sine?

In the previous lesson, we saw that a periodic model is appropriate for a variety of phenomena. This lesson introduces you to an equation involving the **sine** function, whose graph is also periodic. (Sine rhymes with "nine.")

1) How can a Ferris wheel be related to the periodic model? Answer in words or with a labeled graph.

Credit: vitalliy/Fotolia

Objectives for the lesson

You will understand that:

- ☐ The sine function is periodic.

You will be able to:

- ☐ Describe the effect that changing one or more parameters has on the graph of a sine function.
- ☐ Change the parameters of the sine curve to match given criteria.

The sine equation can be used to model smooth periodic behavior, such as the tides and day length. The shape of the curve depends on four parameters. You will investigate the effect that each parameter has on the shape of the curve.

Open Spreadsheet_12.B_Investigating_the_Sine_Curve. The worksheet contains a graph similar in shape to the graphs in the last lesson. There are four parameters to investigate: *A*, *B*, *C*, and *D*. Look at their location in the equation. Also notice that **sin** is used as an abbreviation for sine in the equation. You still pronounce it as a rhyme with "nine."

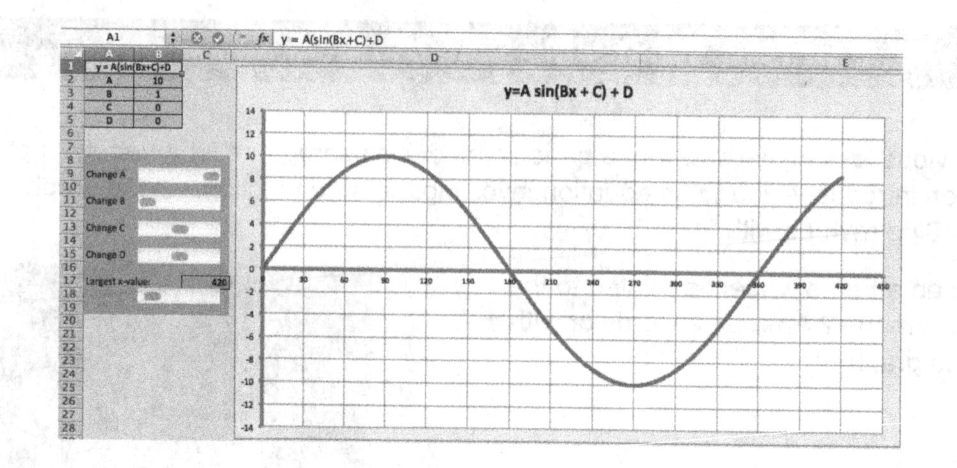

2) Consider the function $y = A sin(Bx + C) + D$:

Part A: What is the input (independent variable)?

Part B: What is the output (dependent variable)?

Part C: What letters represent parameters?

3) Since you already know periodic curves have a period and amplitude, let's find out which parameters control those attributes of the sine curve and how the other parameters change the shape of the curve.

Part A: Use the scroll bar to change the value of A. Examine the graph as you do so to determine the effect of the value of A on the shape of the graph.

Part B: Use the scroll bar to change the value of B. Examine the graph as you do so to determine the effect of the value of B on the shape of the graph.

Part C: Use the scroll bar to change the value of C. Examine the graph as you do so to determine the effect of the value of C on the shape of the graph.

Part D: Use the scroll bar to change the value of D. Examine the graph as you do so to determine the effect of the value of D on the shape of the graph.

4) Now click on the tab for the worksheet ***SIN( )***. Recall from the Preview Assignment that the argument is the expression inside the parentheses. The argument for a sine function can be a value that represents the degree measure of an angle.

Part A: Use the sheet ***SIN( )*** to complete the table. The first row is done for you. Recall that Δ represents change and, in this case, is the change in the curve from $sin(x)$ to $sin(x+1)$.

x (degrees)	$sin(x)$	$sin(x+1)$	Δ (column 3 – column 2)
0	$\sin(0) =$ 0.0000	$\sin(0+1) =$ $\sin(1) =$ .0175	0.0175
15	$\sin(15) =$ 0.2588	$\sin(15+1) =$ $\sin(16) =$?	
30	0.5000		
45		0.7193	
60	0.8660		
75	0.9659		
89		1.0000	

Part B: As you noticed in the last activity, periodic curves change value at different rates. That is, some parts of the curve increase (or decrease) more rapidly than other parts of the curve. For what values of x does the sine curve change relatively rapidly? When is it changing more slowly?

5) Return to the first tab, ***SINE_Curve***. What would the values of the parameters need to be in $y = Asin(Bx + C) + D$ in order to generate the graph of $y = \sin(x)$?

$A =$ $B =$ $C =$ $D =$

6) Change the parameters as indicated in question 5 to see the graph of $y=sin(x)$. Be ready to show how the graph you see supports the findings from the table in question 4.

7) Classify each scenario as most likely to be periodic, exponential, logistic, or linear.

Part A: The amount of electricity used each day by a street lamp that comes on at dusk and turns off at dawn.

Part B: The amount of electricity used by a light that stays on continuously.

Part C: The water level on a pylon (pole) for a pier in the Gulf of Mexico near New Orleans.

Part D: The water temperature measured each day throughout the year in the Gulf of Mexico near New Orleans.

Lesson 12, Part C SIR Disease

In Preview Assignment 12.C, you read about death, disease, and SIR models. In this lesson, you will build a time series model to describe the spread of a disease through a population.

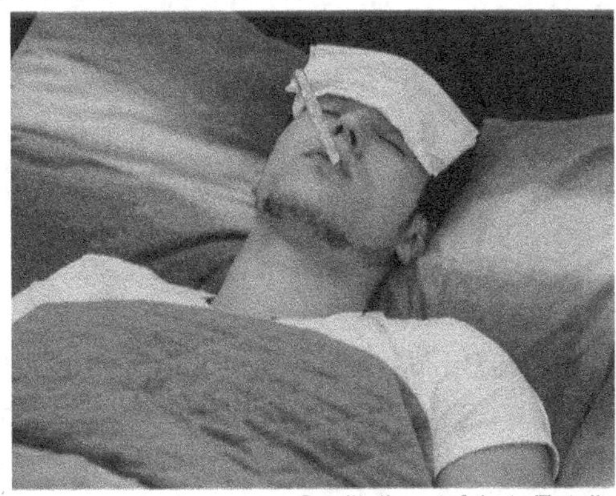

Credit: theartofphoto/Fotolia

1) Which was more deadly, the European outbreak of the Black Death in the 14th century or the flu pandemic of 1918–1919?

Objectives for the lesson

You will begin to understand:

- ☐ What a Susceptible-Infected-Recovered (SIR) model is and how it is used to model epidemics.

You will be able to:

- ☐ Calculate the transmission and recovery rates in a SIR model.
- ☐ Determine whether the compartments of a SIR model are increasing or decreasing.
- ☐ Create a time series SIR model using a spreadsheet.

A SIR model of an epidemic treats people as being in one of three groups: those **S**usceptible to the disease, those **I**nfected with the disease, and those **R**ecovered from the disease (and having developed some immunity). We will use the letters S, I, and R to indicate the number of Susceptible, Infected, and Recovered people. We will also use the Greek letter delta, Δ, to represent "the change" in a quantity. For example, ΔS will represent "the change in S" or the change in the number of Susceptible individuals.

These compartment groups are often represented as boxes in a diagram in which arrows are used to indicate possible transitions from one group to another. For the simplest model, Susceptible people become infected and Infected people recover.

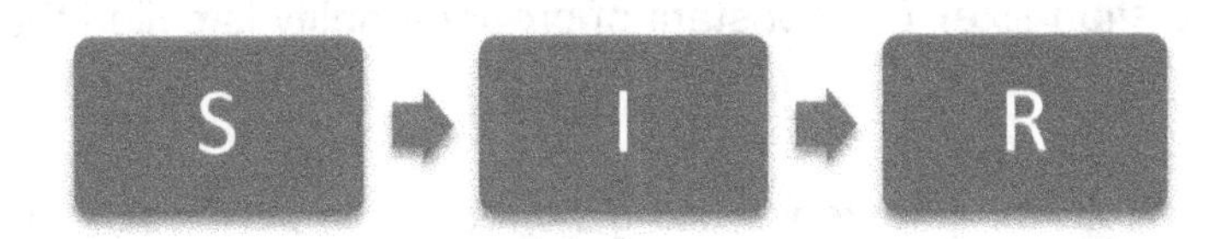

2) The SIR model assumes that the change in the infected population size is jointly proportional to the Susceptible population size and the Infected population size. The constant of proportionality is called the **transmission rate**.

Part A: Write the following sentence using symbols: "The change in the Infected population size is jointly proportional to the Susceptible population size and the Infected population size."

Part B: Assume that if there are 10,000 Susceptible people and 100 Infected people, then 5 Susceptible people will become infected at the next step. Use this information and your formula from Part A to calculate the transmission rate t.

Part C: If there are S susceptible people and I infected people in this step, what is the change in the number of Susceptible people, ΔS?

3) Describe the change in the number of Susceptible people, ΔS.

Part A: When will S increase and when will it decrease?

Part B: How rapidly will S change at the start of the epidemic when I is very small?

Part C: How rapidly will S change at the end of the epidemic when S and I are very small?

Part D: What would the graph of S look like?

4) The SIR model assumes that the change in the Recovered population size is proportional to the Infected population size. The constant of proportionality is called the **recovery rate**, r.

Part A: Write the following sentence using symbols: "The change in the Recovered population size is proportional to the Infected population size."

Part B: Assume that if there are 100 Infected people, then 2 people will recover at the next stage. Use this information and your formula from Part A to calculate the recovery rate.

Part C: By how much will R change from one step to the next?

5) Describe the change in the number of Recovered people, ΔR.

Part A: When will R increase and when will it decrease?

Part B: How rapidly will R change at the start of the epidemic when I is very small?

Part C: How rapidly will R change at the end of the epidemic when I is very small?

Part D: What would the graph of R look like?

6) The only changes to the number of Infected people are increases due to Susceptible people becoming infected (from question 2) and decreases due to Infected people recovering (from question 4). Under the assumptions in questions 3 and 4, what is the change in the number of Infected people, ΔI?

7) Describe the change in the number of Infected people, ΔI.

Part A: When will I increase and when will it decrease?

Part B: How rapidly will I change at the start of the epidemic when I and R are small?

Part C: How rapidly will I change at the end of the epidemic when I and R are small?

Part D: What would the graph of I look like?

Lesson 12, Part D SIR (Continued)

In the last lesson, you developed a basic SIR (Susceptible-Infected-Recovered) model for the transmission of a disease.

1) How could the SIR model be changed to better model diseases or to model other phenomena?

Credit: BlueSkyImages/Fotolia

Objectives for the lesson

You will understand:

- ☐ How you can use mathematics to model phenomena and interventions before they happen or are implemented.
- ☐ How parameters affect the results of a model.

You will be able to:

- ☐ Create a time series SIR model using a spreadsheet.

In this lesson, you will build your own SIR model. Your instructor will give you more information.

2) Begin the planning stages of creating your model.

Part A: Create a diagram showing the phenomenon that your group is going to model.

Part B: With your group, decide on some reasonable assumptions to determine how the individuals in your model move between compartments in the SIR model. What are the parameters of your model?

Part C: List some possible interventions that could be included to improve your model. How will they change the parameters in your model and how do you predict that this change will affect the behavior of the model?

3) Write a short paragraph describing the situation you are going to model. Be sure to describe the parameters in a way that someone new to the project would understand. Include a description of how individuals transition through the different compartments in your model, based on the situation. You will need to explain what you mean by terms such as *transmission rate* and should provide some indication of what these numbers might mean. (For an example, see how transmission rate was explained in In-Class Activities pages, Lesson 12, Part C, question 2.)

4) Use the template spreadsheet and fill in the parameters of your model. Try to select parameters that seem reasonable rather than looking for parameters that simply trivialize the model. Create a line plot to communicate the results of your model.

5) Describe an intervention program that would reduce the severity of the phenomenon that you modeled.

   Part A: Discuss how this intervention would change the parameters.

   Part B: Use the spreadsheet to model the results of this intervention. Create a line plot to communicate the results of the new model.

   Part C: Compare and contrast the line plots to see if the intervention performed as expected.

6) Describe any deficiencies of your model. What other factors that might be involved in the situation are missing in your model?

Lesson 13, Part A Mind the Gap in Income Inequality

In an earlier lesson, we discussed wealth distribution. In recent years, the term *income inequality* has been used more and more. Income inequality can be defined as the difference between the average incomes of the wealthy and the poor.

Economists tell us that the income gap is increasing in most countries. This statement is easy for them to make, but how do they know?

Credit: adrian_ilie825/Fotolia

1) Suppose you are an economist and that you are given the task of measuring income differences between the rich and poor all over the world. Can you survey everyone in the world? How would you proceed to measure income differences?

Objectives for the lesson

You will understand that:

- ☐ A **sample** is a subset of a population.
- ☐ A **statistic** is a numerical summary from a sample.
- ☐ Most **statistical studies** use representative samples to make **inferences** about populations.
- ☐ Some statistical variables are called **explanatory**, while others are called **response;** some are **quantitative** while others are **categorical**.

You will begin to be able to:

- ☐ Describe how a statistical study uses sample data to make inferences about a population.
- ☐ Describe how to gather a representative sample used in a statistical study.
- ☐ Describe the most appropriate statistics to compute in a statistical study.
- ☐ Distinguish between explanatory and response variables, and between quantitative and categorical variables.

It is not possible to measure the incomes of everyone on Earth. It *is* possible to collect information from a smaller group of diverse people from countries all over the world.

(Remember that we would need to convert incomes to a common unit of currency to make comparisons.)

2) What do we call the whole group that we are interested in? What do we call the representative group that we actually collect information from?

3) Recall earlier discussions about census data and income levels in the United States. Now let's design a statistical study about the income of persons from the entire world.

Part A: How would you gather a sample of people so that their incomes are representative of those of the world's entire population? Describe a process that researchers could use.

Part B: To measure income inequality, we decide to divide the sample into three economic groups: lower class, middle class, and upper class. These groups are defined based on their ability to provide themselves with basic needs (food, clothing, shelter) and luxury items. How could we summarize the incomes in each of the three groups? Describe a mathematical process that would provide a simple numerical summary value for the income levels in each group.

Part C: How could we confirm or refute the economists' claims that the income gap is widening?

4) There are two types of studies that researchers conduct by gathering data from population members. In a **census**, data are gathered from an entire population. In a **statistical study**, a representative sample is gathered and results are used to make **inferences** (conclusions based on evidence) about entire populations. Summary values computed from sample data are called **statistic(s)**.

Part A: Is our study a *census* or a *statistical study*? Explain your answer.

Part B: What statistic(s) is/are being computed in our study?

5) When conducting a statistical study, observations that change from one sample member to the next are called **variables**. In this study, two variables are considered.

   Part A: What are the two variables in the study?

   Part B: Classify each variable as quantitative or categorical.

6) There are two primary types of **summary values**: percentages (or proportions) and averages.

   Part A: Which summary value is most appropriate for a quantitative variable, a percentage or an average?

   Part B: Which summary value is most appropriate for a categorical variable, a percentage or an average?

7) Sometimes when two variables are measured in a study, one is called the **explanatory variable**, while the other is called the **response variable**. The researchers may think that values of the explanatory variable might partially explain the differences in the values observed in the response variable. (Caution: This possibility does not imply a cause-and-effect relationship.)

   Part A: Of the two variables in this study, which is the explanatory variable?

   Part B: Which is the response variable?

   Part C: Why do you feel that the explanatory variable might explain the response?

8) Is there any chance that a statistical study such as this can lead to incorrect **inferences** about the entire population? Explain your answer.

✓ **Below are key vocabulary terms used in this lesson for your statistics dictionary. Note that the terms are not alphabetized because they are placed with their associated terms (such as quantitative and categorical variables).**

- **Population** – an entire collection of individuals of interest
- **Sample** – a subset of a population
- **Data** – plural for datum, pieces of information
- **Census** – a study of an entire population
- **Statistical study** – a study that uses sample data to make inferences about populations
- **Statistical inference** – a conclusion based on evidence
- **Statistic** – a summary value computed from sample data
- **Variable** – observations that change among sample (and population) members
- **Quantitative variable** – a variable whose values are quantities
- **Categorical variable** – a variable whose values are non-quantities
- **Response variable** – a particular quantity that we ask a question about in a study
- **Explanatory variable** – a variable that may explain (to some degree) changes observed in another variable, the response variable, a factor that may influence the response variable. The terms *response* and *explanatory* do not imply cause and effect.

Lesson 13, Part B When in Rome . . .

The traditional diet of Mediterranean people (in southern Europe) has generated a lot of interest in recent years.

1) What are some features that distinguish a Mediterranean diet from a typical American diet?

Credit: Gorilla/Fotolia

2) You read about two types of statistical studies in Preview Assignment 13.B: *observational studies* and *experimental studies*. What distinguishes an experimental study from an observational study?

Objectives for the lesson

You will understand that:

- ☐ There are two types of statistical studies: observational and experimental.
- ☐ In observational studies, data are used to make inferences about a population.
- ☐ In experiments, data are used to make inferences about a treatment.

You will be able to:

- ☐ Determine whether a statistical study is observational or experimental.
- ☐ Make appropriate conclusions from observational and experimental studies.

3) Let's extend your understanding of the two types of statistical studies by analyzing two statistical studies related to the Mediterranean diet.

Study X[1]: This study randomly selected men from various countries around the world and collected information about diet and health. The men who ate Mediterranean diets had low rates of heart disease when compared to the other men in the study.

---

[1] The Seven Countries Study. Retrieved October 13, 2014, from www.sevencountriesstudy.com/about-the-study.

Study Y[2]: This study gathered more than 7,000 people in Spain who were at high risk of heart disease. Participants were randomly assigned to one of three groups:

- Mediterranean diet supplemented with extra-virgin olive oil
- Mediterranean diet supplemented with mixed nuts
- Control group (advised to reduce dietary fat intake)

Participants were followed for nearly five years. The groups on the Mediterranean diet had significantly lower rates of heart disease.

Part A: Which study (X or Y) is observational and which is experimental? How do you know?

Part B: A variable is a piece of data that is recorded for each member of a sample. Each of the studies above considers the same two variables. What two variables are observed in these studies?

Part C: One of the variables you listed in Part B is explanatory and the other is response. Which variable is which?

Part D: Are the variables above categorical or quantitative?

Part E: What treatment is applied in the experimental study?

4) Let's look at observational and experimental studies in another way. In an observational study, researchers gather data in representative samples to make inferences (typically) about a population. In an experimental study, researchers gather data to make inferences about a treatment.

Part A: More specifically, in an observational study, researchers look for relationships between variables in a specific population. They may mention the association but do not conclude that another variable might be related to factors causing the

[2] Estruch, R., Ros, E., Salas-Salvadó, J., Covas, M., Corella, D., Arós, F. . . . Martínez-González, M. A. (2013). Primary prevention of cardiovascular disease with a Mediterranean diet. *The New England Journal of Medicine, 368, 1279–1290.*

difference. Either way, remember that the conclusions are statistical inferences that are probably true but are sometimes incorrect.

Write a conclusion to the observational study that describes the relationship observed between the explanatory and response variables.

Part B: In an experimental study, researchers manipulate the outcomes of the explanatory variable among similar groups. If they see significant differences in the response variable between the groups, they will conclude that the varying treatments are the cause.

Write a conclusion to the experimental study that describes the relationship discovered between the explanatory and response variables.

Lesson 13, Part C A Lesson Worth Weighting For

The administration at your community college is concerned about a report that the mean body mass index (BMI) for U.S. adults is increasing. Administrators are considering a college health awareness campaign, but first, they want a measure of students' average BMI.

1) There are too many students at your college to measure the BMI for everyone.

   Part A: Describe a process that could help your college estimate the mean BMI for their student population.

   Part B: Which would be more appropriate, an observational or an experimental study?

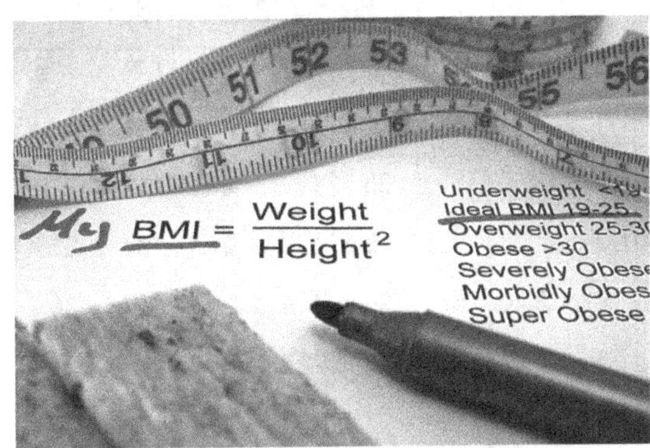

Credit: designer491/Fotolia

Objectives for the lesson

You will understand that:

- ☐ In observational studies, representative samples are used to make inferences about populations.
- ☐ Simple random sampling (SRS) can generate representative samples.
- ☐ Representative samples yield statistics that are similar to population parameters.
- ☐ Statistics from samples always include sampling error; larger samples tend to yield statistics with smaller error.

You will begin to be able to:

- ☐ Identify the principles that would generate a representative sample.
- ☐ Implement a sampling process for generating a representative sample.

We use representative samples in observational studies. One way to gather a representative sample is through **simple random selection**. When gathering data through random selection, every member of the population has the same chance of being selected. A **random sample** is one that is chosen by random selection.

An observational study will help us learn about the mean BMI at the college. Because there are too many students at your college to measure each of their BMI values, you decide to identify 20 students and compute a sample mean BMI.

Let's assume that your college has 540 students. Each of these students has a BMI that is unknown to your college. The following table assigns an ID number (ID#) to each student (from 1 to 540) and gives the respective student's BMI.

Body Mass Index for 540 Community College Students

ID#	BMI	ID#	BMI	ID#	BMI	ID#	BMI	#	BMI	ID#	BMI	ID#	BMI	ID#	BMI	ID#	BMI
1	29.1	61	21.4	121	30.6	181	28.7	241	29.9	301	26.9	361	24.8	421	24.9	481	27.0
2	29.0	62	25.5	122	27.2	182	27.8	242	22.5	302	27.7	362	25.8	422	24.4	482	28.4
3	23.6	63	33.5	123	25.1	183	26.4	243	24.7	303	30.2	363	29.5	423	23.2	483	33.1
4	23.6	64	25.8	124	25.3	184	30.2	244	30.3	304	26.2	364	23.6	424	34.3	484	28.6
5	23.6	65	23.7	125	26.3	185	27.4	245	29.0	305	29.9	365	30.9	425	29.8	485	23.7
6	20.2	66	27.4	126	30.1	186	26.4	246	26.4	306	28.7	366	27.7	426	22.4	486	28.1
7	27.1	67	28.5	127	30.7	187	28.7	247	29.6	307	29.5	367	25.0	427	27.9	487	26.0
8	24.4	68	31.7	128	25.7	188	25.9	248	27.4	308	22.5	368	20.9	428	25.4	488	25.7
9	27.8	69	27.0	129	25.4	189	28.1	249	22.6	309	30.8	369	27.5	429	32.1	489	25.9
10	30.8	70	23.8	130	25.7	190	33.0	250	27.9	310	26.0	370	28.6	430	26.7	490	29.6
11	27.5	71	26.8	131	31.3	191	29.2	251	27.4	311	23.6	371	27.9	431	26.7	491	30.2
12	27.1	72	25.1	132	28.4	192	26.8	252	22.1	312	28.3	372	25.9	432	24.0	492	25.8
13	27.9	73	27.5	133	27.6	193	25.0	253	26.0	313	28.3	373	24.0	433	26.2	493	25.5
14	30.4	74	28.7	134	25.6	194	25.7	254	28.9	314	29.6	374	24.8	434	28.2	494	27.1
15	26.4	75	27.5	135	19.7	195	26.5	255	29.4	315	20.4	375	24.3	435	26.2	495	27.3
16	31.0	76	29.7	136	28.4	196	29.7	256	22.2	316	26.1	376	28.8	436	29.7	496	23.7
17	21.0	77	24.9	137	29.4	197	29.2	257	26.9	317	27.7	377	23.5	437	24.9	497	28.5
18	24.4	78	26.0	138	29.0	198	32.6	258	28.3	318	29.5	378	25.5	438	25.9	498	27.7
19	29.0	79	24.9	139	29.3	199	25.6	259	26.9	319	28.5	379	25.6	439	26.0	499	25.1
20	29.5	80	24.6	140	29.1	200	23.6	260	29.6	320	30.6	380	25.6	440	25.8	500	24.9
21	23.8	81	24.7	141	25.1	201	30.2	261	28.7	321	25.4	381	24.5	441	32.8	501	19.2
22	28.3	82	27.2	142	27.0	202	26.5	262	27.1	322	24.5	382	30.7	442	30.4	502	30.5
23	27.1	83	26.1	143	22.4	203	25.5	263	19.9	323	24.9	383	21.0	443	29.2	503	25.2
24	33.2	84	26.0	144	29.4	204	31.8	264	26.9	324	26.6	384	24.2	444	25.1	504	22.3
25	30.6	85	19.9	145	29.7	205	21.0	265	28.4	325	25.2	385	26.5	445	19.6	505	31.8
26	25.8	86	27.4	146	25.3	206	27.4	266	31.1	326	26.9	386	27.2	446	27.8	506	27.2

27 : 26.4	87 : 23.8	147 : 26.1	207 : 29.7	267 : 25.2	327 : 34.8	387 : 29.9	447 : 27.5	507 : 29.9
28 : 24.9	88 : 28.2	148 : 26.5	208 : 27.9	268 : 26.4	328 : 29.9	388 : 29.7	448 : 21.6	508 : 31.3
29 : 25.6	89 : 24.9	149 : 29.5	209 : 27.9	269 : 23.3	329 : 27.4	389 : 25.4	449 : 30.6	509 : 21.3
30 : 23.8	90 : 24.2	150 : 28.6	210 : 26.9	270 : 27.5	330 : 27.1	390 : 30.7	450 : 26.9	510 : 31.2
31 : 27.1	91 : 29.3	151 : 24.1	211 : 23.6	271 : 29.5	331 : 28.9	391 : 27.3	451 : 27.0	511 : 27.9
32 : 25.6	92 : 29.3	152 : 32.2	212 : 24.6	272 : 32.2	332 : 28.8	392 : 24.3	452 : 25.8	512 : 24.9
33 : 26.0	93 : 28.1	153 : 26.6	213 : 26.8	273 : 22.0	333 : 30.8	393 : 27.2	453 : 26.5	513 : 25.5
34 : 32.3	94 : 35.3	154 : 30.0	214 : 27.2	274 : 25.9	334 : 29.3	394 : 28.4	454 : 31.3	514 : 27.4
35 : 29.1	95 : 23.9	155 : 25.0	215 : 27.1	275 : 23.0	335 : 20.7	395 : 27.5	455 : 22.4	515 : 22.3
36 : 30.2	96 : 26.8	156 : 26.1	216 : 30.5	276 : 25.1	336 : 30.0	396 : 27.0	456 : 23.0	516 : 28.5
37 : 27.6	97 : 27.5	157 : 28.3	217 : 27.5	277 : 25.1	337 : 21.7	397 : 27.2	457 : 29.8	517 : 29.8
38 : 31.2	98 : 23.1	158 : 23.9	218 : 25.2	278 : 28.1	338 : 27.7	398 : 24.9	458 : 25.3	518 : 24.9
39 : 26.0	99 : 26.7	159 : 27.0	219 : 21.8	279 : 28.6	339 : 24.6	399 : 26.7	459 : 23.8	519 : 23.7
40 : 25.9	100 : 30.2	160 : 24.1	220 : 27.3	280 : 22.9	340 : 23.9	400 : 24.5	460 : 29.1	520 : 25.9
41 : 31.1	101 : 22.4	161 : 24.5	221 : 33.0	281 : 29.8	341 : 30.9	401 : 29.0	461 : 28.8	521 : 29.2
42 : 25.7	102 : 25.6	162 : 28.3	222 : 27.4	282 : 29.1	342 : 25.7	402 : 24.1	462 : 20.9	522 : 29.3
43 : 29.7	103 : 25.0	163 : 27.2	223 : 25.5	283 : 27.9	343 : 27.7	403 : 24.9	463 : 25.6	523 : 23.8
44 : 25.5	104 : 32.6	164 : 24.9	224 : 26.9	284 : 28.5	344 : 32.0	404 : 30.5	464 : 33.1	524 : 23.4
45 : 20.4	105 : 26.6	165 : 26.7	225 : 24.9	285 : 26.5	345 : 25.0	405 : 28.3	465 : 31.2	525 : 27.6
46 : 28.8	106 : 29.5	166 : 25.9	226 : 23.0	286 : 25.0	346 : 27.1	406 : 31.2	466 : 34.0	526 : 30.9
47 : 26.9	107 : 29.3	167 : 23.5	227 : 23.8	287 : 28.1	347 : 31.1	407 : 29.1	467 : 31.7	527 : 28.8
48 : 29.7	108 : 29.7	168 : 25.4	228 : 24.0	288 : 31.5	348 : 24.4	408 : 27.6	468 : 28.0	528 : 29.5
49 : 28.9	109 : 25.8	169 : 30.1	229 : 26.8	289 : 23.7	349 : 29.9	409 : 29.8	469 : 29.7	529 : 24.0
50 : 31.6	110 : 27.9	170 : 27.6	230 : 24.4	290 : 23.2	350 : 27.8	410 : 27.4	470 : 30.3	530 : 30.0
51 : 24.9	111 : 23.6	171 : 24.9	231 : 24.6	291 : 25.2	351 : 24.9	411 : 26.0	471 : 28.0	531 : 32.7
52 : 28.7	112 : 26.0	172 : 23.2	232 : 24.8	292 : 27.4	352 : 29.9	412 : 24.8	472 : 30.5	532 : 28.4
53 : 26.2	113 : 25.7	173 : 27.8	233 : 30.8	293 : 31.3	353 : 26.6	413 : 21.0	473 : 31.2	533 : 27.0
54 : 31.1	114 : 23.0	174 : 24.6	234 : 27.1	294 : 26.3	354 : 27.6	414 : 24.8	474 : 29.3	534 : 25.4
55 : 22.1	115 : 21.3	175 : 26.5	235 : 23.6	295 : 21.1	355 : 24.9	415 : 27.4	475 : 24.4	535 : 31.7
56 : 25.1	116 : 30.8	176 : 25.9	236 : 29.6	296 : 31.4	356 : 28.6	416 : 28.6	476 : 27.2	536 : 26.3
57 : 20.9	117 : 29.7	177 : 27.7	237 : 26.6	297 : 30.0	357 : 23.7	417 : 30.5	477 : 29.6	537 : 31.9
58 : 26.1	118 : 31.2	178 : 32.9	238 : 35.2	298 : 25.7	358 : 28.5	418 : 25.2	478 : 31.5	538 : 28.2
59 : 27.3	119 : 24.5	179 : 29.2	239 : 25.9	299 : 24.1	359 : 26.8	419 : 28.7	479 : 29.7	539 : 27.6
60 : 17.1	120 : 23.0	180 : 28.9	240 : 21.5	300 : 22.8	360 : 28.6	420 : 26.9	480 : 24.1	540 : 27.6

While this population of 540 is small in comparison to other populations, it is still too large for the administration to survey every student. Measuring a student BMI requires a visit to the college's health center, and 540 visits are not possible given time and staff constraints.

2) One way to randomly select members from this population is to use a **random number generator**. Many calculators and computer programs have random number generators.

   Part A: Randomly select 20 student ID numbers. If using a random number table, pick an arbitrary starting point, and write that number on the left of the first line of your notebook. From your starting point, move right or left in the table, recording the ID numbers in a column down your notebook paper.

   Part B: Locate the 20 IDs in the Body Mass Index table. Write the corresponding BMI next to each ID number.

   Part C: Now compute the mean BMI for your random sample of students.

3) A **statistic** is a summary value from a sample. One example of a statistic is the sample mean you just computed.

   A summary value from an entire population is called a **parameter**. One example of a parameter is a population mean (if you were to measure the BMI of all students and compute the mean).

   Part A: Consider the BMI for your sample of twenty students and compare those values to the BMI values of the population. Does the sample appear to be representative of the population?

   Part B: If you believe your sample is representative of the population, you can use the mean to make a statistical inference about the mean BMI for the population. Write an inference statement about the population.

Part C: A person with a BMI greater than or equal to 25 is often considered overweight. Does the evidence you gathered lead you to conclude that the college administration should implement the health awareness campaign?

4) The mean BMI for the population of all 540 college students is 27.0.

Part A: Did the statistic computed from your sample come close to this value?

Part B: The **absolute error** in a statistic is equal to the absolute value of the difference between the statistic and the corresponding population parameter (in this case, the *population mean*). What is the absolute error in the mean computed from the random sample of size 20?

Part C: What is the cause of this error?

5) Is this research an observational or experimental study? How do you know?

In the next lesson, we will consider another way of gathering a representative sample that will result in statistics with errors as well. We will compare the mean error in that lesson with the mean error you just computed. This comparison will help us determine whether that method works better than random sampling.

Lesson 13, Part D Weight . . . There's More!

In this lesson, we continue to improve our design of an observational study that your college is planning to conduct about the average body mass index (BMI) of its students. Since the last lesson, two important facts have come to the attention of the administrators of the study.

- 60% of students at the college are female.
- Females have a lower BMI on average.

These facts are causing them to reconsider how to gather data. The college administrators are concerned that a simple random sample may not be representative of the student population.

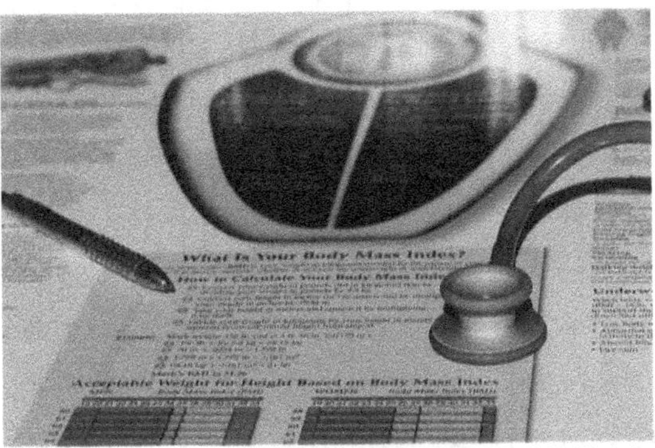

Credit: Keith Frith/Fotolia

1) How could a representative sample of students be collected that addresses the additional concerns?

Objectives for the lesson

You will understand that:

- ☐ In observational studies, representative samples are used to make inferences about populations.
- ☐ Stratified sampling can generate representative samples.
- ☐ Stratified samples tend to yield statistics with smaller errors than random samples.

You will be able to:

- ☐ Create and implement a stratified sampling process for generating a representative sample.

Simple random sampling is often used to generate a representative sample to address a research question or hypothesis for an observational study. **Stratified sampling** is another method often used when the overall population has subpopulations that may have different characteristics. With stratified sampling, researchers divide the members of the population into its homogeneous, logical, appropriate subgroups/subsets. Random samples are taken from each group, according to each group's percentage of the overall population.

In our observational study about BMI, on average, females tend to have lower BMI than males. Therefore, it makes sense to make sure that the population is divided into females and males to be sure that each subgroup is represented properly in our sample. Another population division that might be important has to do with age. Young adults have lower BMIs, on average, than older adults.

Body Mass Index for 540 Community College Students

#ID: BMI (Gender/Age)

#1:29.1 (F/23)	#61:21.4 (F/30)	#121:30.6 (F/36)	#181:28.7 (M/19)	#241:29.9 (F/28)	#301:26.9 (F/28)	#361:24.8 (F/26)	#421:24.9 (F/31)	#481:27.0 (M/21)
#2:29.0 (F/22)	#62:25.5 (F/20)	#122:27.2 (F/18)	#182:27.8 (F/29)	#242:22.5 (F/17)	#302:27.7 (F/27)	#362:25.8 (F/29)	#422:24.4 (M/26)	#482:28.4 (F/34)
#3:23.6 (F/19)	#63:33.5 (F/51)	#123:25.1 (F/20)	#183:26.4 (F/23)	#243:24.7 (F/31)	#303:30.2 (M/25)	#363:29.5 (M/25)	#423:23.2 (M/17)	#483:33.1 (M/26)
#4:23.6 (F/23)	#64:25.8 (F/20)	#124:25.3 (F/21)	#184:30.2 (F/44)	#244:30.3 (M/25)	#304:26.2 (M/20)	#364:23.6 (M/26)	#424:34.3 (M/26)	#484:28.6 (M/25)
#5:23.6 (F/27)	#65:23.7 (F/26)	#125:26.3 (F/20)	#185:27.4 (F/19)	#245:29.0 (M/18)	#305:29.9 (F/29)	#365:30.9 (F/46)	#425:29.8 (M/24)	#485:23.7 (M/21)
#6:20.2 (F/29)	#66:27.4 (F/17)	#126:30.1 (M/25)	#186:26.4 (M/19)	#246:26.4 (M/26)	#306:28.7 (M/30)	#366:27.7 (M/26)	#426:22.4 (M/23)	#486:28.1 (M/21)
#7:27.1 (F/17)	#67:28.5 (F/22)	#127:30.7 (F/41)	#187:28.7 (M/19)	#247:29.6 (F/38)	#307:29.5 (M/20)	#367:25.0 (M/24)	#427:27.9 (F/31)	#487:26.0 (M/27)
#8:24.4 (F/18)	#68:31.7 (F/53)	#128:25.7 (M/17)	#188:25.9 (F/27)	#248:27.4 (F/24)	#308:22.5 (F/22)	#368:20.9 (F/29)	#428:25.4 (M/23)	#488:25.7 (M/27)
#9:27.8 (F/18)	#69:27.0 (F/17)	#129:25.4 (F/19)	#189:28.1 (F/20)	#249:22.6 (F/17)	#309:30.8 (M/25)	#369:27.5 (M/18)	#429:32.1 (M/48)	#489:25.9 (M/24)
#10:30.8 (F/38)	#70:23.8 (F/29)	#130:25.7 (M/19)	#190:33.0 (M/25)	#250:27.9 (F/25)	#310:26.0 (F/32)	#370:28.6 (M/20)	#430:26.7 (M/26)	#490:29.6 (M/25)
#11:27.5 (F/18)	#71:26.8 (F/18)	#131:31.3 (F/49)	#191:29.2 (M/23)	#251:27.4 (F/28)	#311:23.6 (M/24)	#371:27.9 (F/30)	#431:26.7 (M/29)	#491:30.2 (M/26)
#12:27.1 (F/20)	#72:25.1 (F/19)	#132:28.4 (F/25)	#192:26.8 (F/23)	#252:22.1 (F/28)	#312:28.3 (F/32)	#372:25.9 (M/19)	#432:24.0 (F/32)	#492:25.8 (M/21)
#13:27.9 (F/19)	#73:27.5 (F/20)	#133:27.6 (F/26)	#193:25.0 (F/31)	#253:26.0 (F/32)	#313:28.3 (M/23)	#373:24.0 (F/24)	#433:26.2 (M/24)	#493:25.5 (M/28)
#14:30.4 (F/38)	#74:28.7 (F/31)	#134:25.6 (M/17)	#194:25.7 (M/21)	#254:28.9 (F/36)	#314:29.6 (M/20)	#374:24.8 (F/26)	#434:28.2 (F/25)	#494:27.1 (M/28)
#15:26.4 (F/17)	#75:27.5 (F/23)	#135:19.7 (F/18)	#195:26.5 (F/31)	#255:29.4 (F/33)	#315:20.4 (F/28)	#375:24.3 (F/17)	#435:26.2 (M/24)	#495:27.3 (M/20)
#16:31.0 (M/24)	#76:29.7 (F/29)	#136:28.4 (F/25)	#196:29.7 (F/34)	#256:22.2 (F/19)	#316:26.1 (M/22)	#376:28.8 (F/26)	#436:29.7 (M/22)	#496:23.7 (M/22)
#17:21.0 (F/24)	#77:24.9 (F/20)	#137:29.4 (F/29)	#197:29.2 (M/26)	#257:26.9 (F/20)	#317:27.7 (M/19)	#377:23.5 (F/27)	#437:24.9 (M/18)	#497:28.5 (M/20)
#18:24.4 (F/31)	#78:26.0 (F/26)	#138:29.0 (F/36)	#198:32.6 (M/52)	#258:28.3 (F/31)	#318:29.5 (M/22)	#378:25.5 (F/31)	#438:25.9 (M/23)	#498:27.7 (M/34)
#19:29.0 (F/33)	#79:24.9 (F/20)	#139:29.3 (F/37)	#199:25.6 (M/18)	#259:26.9 (F/22)	#319:28.5 (F/25)	#379:25.6 (M/20)	#439:26.0 (M/23)	#499:25.1 (F/31)
#20:29.5 (F/36)	#80:24.6 (F/18)	#140:29.1 (F/30)	#200:23.6 (F/28)	#260:29.6 (M/20)	#320:30.6 (F/39)	#380:25.6 (M/21)	#440:25.8 (M/17)	#500:24.9 (M/22)
#21:23.8 (F/19)	#81:24.7 (F/22)	#141:25.1 (F/21)	#201:30.2 (M/25)	#261:28.7 (M/21)	#321:25.4 (F/30)	#381:24.5 (M/31)	#441:32.8 (M/50)	#501:19.2 (F/18)
#22:28.3 (F/21)	#82:27.2 (F/17)	#142:27.0 (F/31)	#202:26.5 (M/20)	#262:27.1 (F/27)	#322:24.5 (M/24)	#382:30.7 (F/50)	#442:30.4 (M/23)	#502:30.5 (M/32)
#23:27.1 (F/21)	#83:26.1 (F/23)	#143:22.4 (F/25)	#203:25.5 (F/25)	#263:19.9 (M/27)	#323:24.9 (F/26)	#383:21.0 (M/21)	#443:29.2 (M/28)	#503:25.2 (M/24)
#24:33.2 (F/48)	#84:26.0 (F/30)	#144:29.4 (F/30)	#204:31.8 (M/25)	#264:26.9 (F/27)	#324:26.6 (M/19)	#384:24.2 (M/23)	#444:25.1 (F/31)	#504:22.3 (F/23)
#25:30.6 (F/33)	#85:19.9 (M/21)	#145:29.7 (F/30)	#205:21.0 (M/21)	#265:28.4 (F/29)	#325:25.2 (F/32)	#385:26.5 (M/24)	#445:19.6 (M/24)	#505:31.8 (M/25)
#26:25.8 (F/19)	#86:27.4 (F/18)	#146:25.3 (F/22)	#206:27.4 (F/19)	#266:31.1 (M/25)	#326:26.9 (F/30)	#386:27.2 (F/27)	#446:27.8 (M/27)	#506:27.2 (M/23)
#27:26.4 (F/21)	#87:23.8 (F/30)	#147:26.1 (F/24)	#207:29.7 (F/36)	#267:25.2 (F/28)	#327:34.8 (M/26)	#387:29.9 (M/26)	#447:27.5 (M/21)	#507:29.9 (M/27)
#28:24.9 (F/17)	#88:28.2 (F/21)	#148:26.5 (F/20)	#208:27.9 (F/21)	#268:26.4 (M/27)	#328:29.9 (F/31)	#388:29.7 (M/20)	#448:21.6 (F/25)	#508:31.3 (M/42)
#29:25.6 (F/22)	#89:24.9 (F/23)	#149:29.5 (F/39)	#209:27.9 (F/21)	#269:23.3 (F/18)	#329:27.4 (F/31)	#389:25.4 (M/19)	#449:30.6 (M/29)	#509:21.3 (M/19)
#30:23.8 (F/24)	#90:24.2 (F/18)	#150:28.6 (F/24)	#210:26.9 (F/20)	#270:27.5 (F/33)	#330:27.1 (F/28)	#390:30.7 (M/24)	#450:26.9 (M/27)	#510:31.2 (M/46)
#31:27.1 (F/22)	#91:29.3 (F/25)	#151:24.1 (F/18)	#211:23.6 (F/32)	#271:29.5 (M/19)	#331:28.9 (M/23)	#391:27.3 (M/18)	#451:27.0 (M/20)	#511:27.9 (M/20)
#32:25.6 (F/27)	#92:29.3 (F/25)	#152:32.2 (F/48)	#212:24.6 (F/29)	#272:32.2 (F/50)	#332:28.8 (F/24)	#392:24.3 (M/19)	#452:25.8 (M/18)	#512:24.9 (M/24)
#33:26.0 (F/19)	#93:28.1 (F/19)	#153:26.6 (F/25)	#213:26.8 (F/30)	#273:22.0 (M/18)	#333:30.8 (M/25)	#393:27.2 (F/30)	#453:26.5 (M/26)	#513:25.5 (M/29)
#34:32.3 (F/57)	#94:35.3 (M/26)	#154:30.0 (F/30)	#214:27.2 (F/26)	#274:25.9 (F/32)	#334:29.3 (M/19)	#394:28.4 (F/33)	#454:31.3 (F/53)	#514:27.4 (M/30)
#35:29.1 (F/24)	#95:23.9 (F/26)	#155:25.0 (F/23)	#215:27.1 (F/25)	#275:23.0 (F/31)	#335:20.7 (M/17)	#395:27.5 (M/21)	#455:22.4 (M/27)	#515:22.3 (F/30)

#36:30.2 (F/37)	#96:26.8 (F/20)	#156:26.1 (F/26)	#216:30.5 (F/35)	#276:25.1 (F/23)	#336:30.0 (F/42)	#396:27.0 (M/19)	#456:23.0 (M/17)	#516:28.5 (M/21)
#37:27.6 (F/19)	#97:27.5 (F/25)	#157:28.3 (F/21)	#217:27.5 (F/33)	#277:25.1 (F/27)	#337:21.7 (F/32)	#397:27.2 (M/19)	#457:29.8 (M/24)	#517:29.8 (M/26)
#38:31.2 (F/40)	#98:23.1 (M/23)	#158:23.9 (F/31)	#218:25.2 (F/18)	#278:28.1 (F/22)	#338:27.7 (M/22)	#398:24.9 (F/29)	#458:25.3 (M/24)	#518:24.9 (M/24)
#39:26.0 (F/21)	#99:26.7 (F/21)	#159:27.0 (M/17)	#219:21.8 (F/28)	#279:28.6 (F/33)	#339:24.6 (M/22)	#399:26.7 (F/31)	#459:23.8 (M/19)	#519:23.7 (M/23)
#40:25.9 (F/19)	#100:30.2 (F/37)	#160:24.1 (F/20)	#220:27.3 (F/19)	#280:22.9 (F/19)	#340:23.9 (F/32)	#400:24.5 (M/32)	#460:29.1 (M/18)	#520:25.9 (M/24)
#41:31.1 (F/37)	#101:22.4 (F/18)	#161:24.5 (M/17)	#221:33.0 (M/54)	#281:29.8 (F/35)	#341:30.9 (F/45)	#401:29.0 (M/21)	#461:28.8 (M/17)	#521:29.2 (M/35)
#42:25.7 (F/19)	#102:25.6 (F/29)	#162:28.3 (F/26)	#222:27.4 (F/22)	#282:29.1 (F/32)	#342:25.7 (M/23)	#402:24.1 (F/27)	#462:20.9 (M/18)	#522:29.3 (M/21)
#43:29.7 (F/29)	#103:25.0 (F/22)	#163:27.2 (F/22)	#223:25.5 (F/29)	#283:27.9 (F/29)	#343:27.7 (M/24)	#403:24.9 (F/30)	#463:25.6 (M/24)	#523:23.8 (M/20)
#44:25.5 (F/18)	#104:32.6(M/26)	#164:24.9 (F/25)	#224:26.9 (F/20)	#284:28.5 (F/24)	#344:32.0 (M/44)	#404:30.5 (F/41)	#464:33.1 (M/26)	#524:23.4 (F/23)
#45:20.4 (F/22)	#105:26.6 (F/18)	#165:26.7 (F/22)	#225:24.9 (F/26)	#285:26.5 (M/21)	#345:25.0 (M/18)	#405:28.3 (M/36)	#465:31.2 (M/39)	#525:27.6 (F/31)
#46:28.8 (F/21)	#106:29.5 (F/37)	#166:25.9 (F/22)	#226:23.0 (F/30)	#286:25.0 (M/17)	#346:27.1 (F/31)	#406:31.2 (F/43)	#466:34.0 (M/51)	#526:30.9 (M/38)
#47:26.9 (F/18)	#107:29.3 (F/27)	#167:23.5 (F/21)	#227:23.8 (F/30)	#287:28.1 (F/29)	#347:31.1 (M/41)	#407:29.1 (F/35)	#467:31.7 (F/58)	#527:28.8 (M/33)
#48:29.7 (F/29)	#108:29.7 (F/30)	#168:25.4 (F/21)	#228:24.0 (F/22)	#288:31.5 (F/48)	#348:24.4 (M/20)	#408:27.6 (F/31)	#468:28.0 (M/17)	#528:29.5 (M/32)
#49:28.9 (F/23)	#109:25.8 (F/25)	#169:30.1 (M/31)	#229:26.8 (M/18)	#289:23.7 (F/30)	#349:29.9 (F/38)	#409:29.8 (F/38)	#469:29.7 (M/37)	#529:24.0 (M/18)
#50:31.6 (F/48)	#110:27.9 (F/21)	#170:27.6 (F/29)	#230:24.4 (M/18)	#290:23.2 (F/25)	#350:27.8 (M/17)	#410:27.4 (F/31)	#470:30.3 (M/32)	#530:30.0 (M/37)
#51:24.9 (F/20)	#111:23.6 (F/28)	#171:24.9 (F/26)	#231:24.6 (F/31)	#291:25.2 (F/31)	#351:24.9 (F/27)	#411:26.0 (M/17)	#471:28.0 (M/22)	#531:32.7 (M/48)
#52:28.7 (F/23)	#112:26.0 (F/30)	#172:23.2 (F/22)	#232:24.8 (F/22)	#292:27.4 (F/30)	#352:29.9 (M/25)	#412:24.8 (F/27)	#472:30.5 (M/25)	#532:28.4 (M/28)
#53:26.2 (M/18)	#113:25.7 (F/30)	#173:27.8 (F/23)	#233:30.8 (M/24)	#293:31.3 (F/53)	#353:26.6 (M/27)	#413:21.0 (M/24)	#473:31.2 (M/41)	#533:27.0 (M/26)
#54:31.1 (M/25)	#114:23.0 (F/22)	#174:24.6 (F/21)	#234:27.1 (F/26)	#294:26.3 (M/17)	#354:27.6 (F/29)	#414:24.8 (F/30)	#474:29.3 (M/19)	#534:25.4 (M/24)
#55:22.1 (F/17)	#115:21.3 (F/28)	#175:26.5 (F/21)	#235:23.6 (M/22)	#295:21.1 (F/25)	#355:24.9 (F/27)	#415:27.4 (M/18)	#475:24.4 (M/28)	#535:31.7 (M/43)
#56:25.1 (F/17)	#116:30.8 (F/40)	#176:25.9 (F/26)	#236:29.6 (F/37)	#296:31.4 (M/40)	#356:28.6 (F/33)	#416:28.6 (M/23)	#476:27.2 (M/23)	#536:26.3 (M/19)
#57:20.9 (F/19)	#117:29.7 (F/30)	#177:27.7 (F/23)	#237:26.6 (F/29)	#297:30.0 (F/36)	#357:23.7 (M/17)	#417:30.5 (F/47)	#477:29.6 (M/21)	#537:31.9 (M/26)
#58:26.1 (F/20)	#118:31.2 (F/41)	#178:32.9 (F/55)	#238:35.2 (M/26)	#298:25.7 (M/22)	#358:28.5 (F/32)	#418:25.2 (M/19)	#478:31.5 (M/25)	#538:28.2 (F/29)
#59:27.3 (F/17)	#119:24.5 (F/17)	#179:29.2 (F/35)	#239:25.9 (F/28)	#299:24.1 (F/23)	#359:26.8 (M/24)	#419:28.7 (M/31)	#479:29.7 (M/38)	#539:27.6 (M/17)
#60:17.1 (F/23)	#120:23.0 (F/26)	#180:28.9 (F/30)	#240:21.5 (F/30)	#300:22.8 (M/26)	#360:28.6 (M/19)	#420:26.9 (F/32)	#480:24.1 (M/20)	#540:27.6 (M/21)

By running an analysis of the school's registration records, we find that the student population consists of the following percentages.

Group	% of Student Body
Females under 30	38%
Females 30 years and older	22%
Males under 30	34%
Males 30 years and older	6%

In the table on the previous page, the entire student body of 540 community college students is listed again, with information about age and gender added into each entry in parentheses.

Consider student ID #61 at the top of the second column of numbers.

#61:21.4 (F/30)

This record tells us that student #61 has a BMI of 21.4, is female (F), and is 30 years old.

2) Once again, you will select a sample of 20 students, this time stratifying to represent the four groups.

Part A: How many of each group should be present in our representative sample of size 20? Round your values appropriately and be sure that the total sample size is 20.

Part B: Create a table in your notes like the one below, but containing enough rows to hold the number of students calculated for each category.

Women < 30	Women ≥ 30	Men < 30	Men ≥ 30

Part C: Pick an arbitrary start point in the Student ID table and circle it. Look up that student's BMI and write it down under the appropriate column heading below. Move to the right in the table and circle that entry. Again, look up the BMI for that student and record it in the appropriate column. Continue until a category column fills. Further data values for that category should be ignored. When every category is filled, you are done!

Part D: Now compute the mean BMI for the random sample of 20 students that you picked.

Part E: A **statistic** is a summary value from a sample. A **parameter** is a summary value from an entire population. Is your answer to Part D a statistic or a parameter?

Part F: The absolute error in a sample mean is the absolute value of the difference between the sample mean and the population mean. The population mean is a BMI of 27.0. Compute the absolute error in your statistic. Record your answer here and on the board in the classroom.

3) In the previous lesson, your group computed a sample mean from a simple random sample.

Part A: What was your mean and absolute error from the previous lesson using simple random sampling? Record your answers here and record your absolute error on the board in the classroom.

Part B: Which mean had a larger error, the one from the simple random sample or the one from the stratified sample?

Part C: Share your absolute errors with the class. What is the average absolute error among all of the groups?

Part D: Is this mean error smaller for the stratified sample or the random sample from the last lesson?

Lesson 14, Part A Blood Pressure and Bias

When blood pressure is measured, two readings are taken. The higher reading is called the systolic pressure, which is the pressure as the heart beats. The second measure is the diastolic pressure, or the pressure between beats.

You may have had your blood pressure measured by an arm cuff. Newer devices have been developed that can measure blood pressure by simply attaching a cuff to a person's finger.

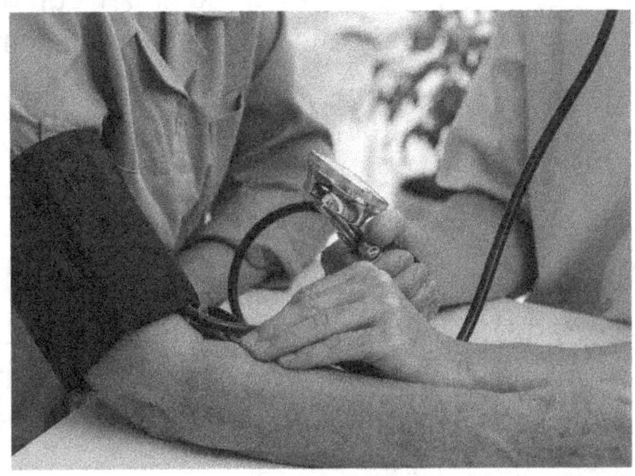

Credit: Photographee.eu/Fotolia

1) What did you learn about blood pressure from your reading in Preview Assignment 14.A?

Objectives for the lesson

You will understand that:

- ☐ Error in an estimate has two components: sampling error and non-sampling error.
- ☐ Non-sampling error arises from sources other than sampling, such as **bias**.

You will be able to:

- ☐ Determine when a sampling process can yield non-sampling errors due to bias.

2) Recall from the Preview Assignment that the blood pressures for thirty adults were measured with the new finger-cuff device (shown in the dotplot, systolic mean = 119.1). Afterward, the pressures were measured again with a reliable inflatable arm-cuff device. The mean systolic pressure was 115.6. *We assume that this is the correct average.*

Part A: Plot this mean with an *X* on the horizontal axis of the dotplot.

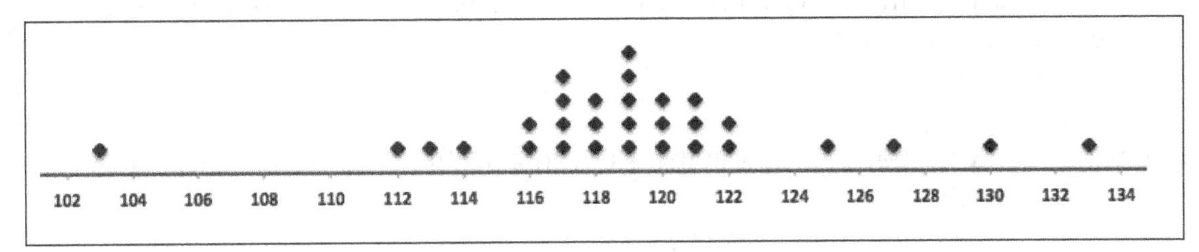

Part B: We assume that the mean of 115.6 is accurate because of the reliability of the arm-cuff device. This difference in results between the two methods implies that the mean of the finger-cuff measurements has some error. The error is the difference between the estimate from the finger-cuff device and the accurate mean. What is the value of the error? Write a contextual sentence about the error.

Part C: Do you think that the distribution of the blood pressures from the finger-cuff device, when compared to the mean value from the arm-cuff device, is due to chance fluctuation or some sort of bias in the measurements?

Part D: Which of the following statements is true? Explain your answer.

a) The blood pressure measurements are all higher than the reliable average.

b) The distribution of blood systolic pressures is shifted to the right, so that the correct mean is off-center.

Part E: In this context, describe the meaning of the word *bias*.

3) We have previously thought about **sampling error**. Sampling error arises when we use a sample instead of the entire population to make an estimate of a population parameter. Error from other sources is called **non-sampling error**. A primary source of non-sampling error is **bias**. Bias occurs when some flaw in the design or implementation of a study results in a tendency for measurements to be too high or too low.

Part A: By the definition of *bias* given above, does it seem that there is a flaw in the design or implementation of this study that is causing measurements to be too high or too low?

Part B: Can you guess the flaw in this study that might cause the bias evident in the data values?

Part C: Can we use the mean systolic blood pressure from the thirty random finger-cuff measurements to make an inference about the mean for all adults?

Part D: Researchers plan to use the finger-cuff device to measure the outcome of a new blood pressure medication that they are studying. How will this affect the doctors' perceptions of the medication's effects?

4) Summarize the key characteristics and important concepts that must be considered when selecting samples and collecting sample data.

Lesson 14, Part B Taking Aim at Bias

According to the U.S. Bureau of Justice Statistics, the proportion of children in grades 9 through 12 who reported having carried a weapon in the previous 30 days is alarmingly high.[1]

Credit: Karen Roach/Fotolia

One researcher conducted a survey to estimate the percentage by gathering a representative sample of children in these grades from across the U.S. Out of 7,435 students surveyed, 1,297 admitted to carrying a weapon in the last month.

1) What percentage of students admitted to carrying a weapon in the last month?

Objectives for the lesson

You will understand that:

- ☐ When sample members fail to give information in a statistical study, **non-response bias** is the result.
- ☐ When the method in which data values are observed or measured affects the values, **response** (or **measurement**) **bias** is the result.
- ☐ When data are gathered in a way that makes some groups of population members less likely to be selected for a sample, **selection bias** is the result.

You will be able to:

- ☐ Identify methods of data gathering where bias is likely to occur.
- ☐ Distinguish the various types of bias in real situations.

In this lesson, we will consider three important sources of bias: **non-response bias**, **selection bias,** and **response** (or **measurement**) **bias**. In Preview Assignment 14.B, you were asked to record these terms and definitions in your notes.

---

[1] Zhang, R. S., & Truman, J. (2012). *Indicators of school crime and safety: 2011* (NCES 2012-002/NCJ 236021). Retrieved November 19, 2014, from http://nces.ed.gov/pubs2012/2012002rev.pdf.

2) The survey described asked several questions about school violence in addition to the question about carrying weapons. An interesting fact about the data from this survey is that 439 students skipped the question on carrying weapons.

Part A: How might having 439 students skip the question impact the conclusions made by the survey?

Part B: Is it possible that people who skipped the question have something in common? Do we know for sure?

Part C: Which type of bias is present if a subgroup of the population declined to participate?

Part D: Think about the true proportion of students who carried weapons in the 30 days prior to the survey. Does the bias you noted in Part C tend to make the survey result of 17% lower or higher than the true proportion? Explain your answer.

3) From the time when women were finally granted the right to vote (in 1920) up until 1980, a higher proportion of men exercised their right to vote when compared with women. However, since 1980, the proportion of women who vote has been greater than men.

Suppose a researcher wishes to learn the proportion of the U.S. voting population that supports stricter gun control. Thinking that women and men voters may have different views on gun control, the researcher decides to gather a stratified sample with 500 female and 500 male voters, randomly selected.

Part A: If the females and males are randomly selected, then we have a good chance that all types of people will be present in the sample. Does this mean that the sample will be representative of the population of U.S. voters?

Part B: **Random selection** means that all population members have the same chance of being selected for a sample. Does any particular woman in the population have the same chance of being selected for this sample? Explain your answer.

Part C: As illustrated in the graphic below, 54% of U.S. voters are women and 46% are men. If we pick the same number of women as men for our sample, is a given woman as **likely** to be chosen as a man?

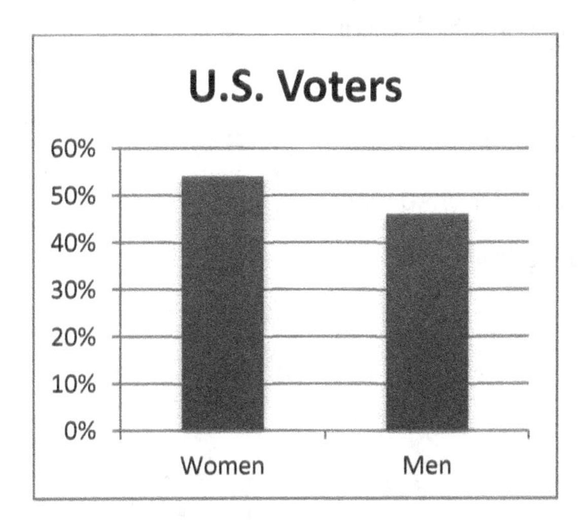

Part D: If the researcher randomly gathers 500 women and 500 men for the survey, the data will be biased. What type of bias will result?

Part E: Typically speaking, a larger proportion of women support stricter gun control.[2] If 500 women and 500 men are selected for the sample, will the proportion of those favoring stricter gun control be higher or lower than it should be?

Part F: Suppose a researcher picks 1,000 voters randomly but is careful to make sure that 540 are women and 460 are men. Will this approach lead to selection bias? Explain your answer.

4) The researcher asked the question on gun violence in the following way: "In 2010, there were approximately 30,000 deaths due to firearms in the U.S. In light of this data, do you feel that stricter gun control is appropriate?"

---

[2] Izadi, E. (2013, April 12). Explaining the gender gap on gun control. *The National Journal Daily.* Retrieved November 19, 2014, from http://www.nationaljournal.com/politics/explaining-the-gender-gap-on-gun-control-20130412.

Part A: How might this question introduce bias?

Part B: What kind of bias is this? Explain.

a) Selection bias

b) Measurement/response bias

c) Non-response bias

Lesson 14, Part C Conclusions in Observational Studies

A college needs to know if students would support a $50 fee increase to improve library facilities. A college researcher gathered 250 students, selected randomly, for the survey using the college database.

The students were asked the following: "The library is a vital part of your education. It is your source for research. Would you support a small fee increase to improve its facilities?" Of those surveyed, 73% supported the fee increase.

Credit: stokkete/Fotolia

1) What is the sample in this case? What is the population?

Objectives for the lesson

You will understand that:

- ☐ Researchers can design their studies to minimize the impact that bias can have over their results.
- ☐ Researchers can make conclusions from observational studies that describe relationships between variables. These conclusions may not include cause and effect explanations.

You will able to:

- ☐ Identify elements of a research design that may introduce bias.
- ☐ Suggest corrections to a research design that can minimize bias.
- ☐ Identify inappropriate conclusions in an observational study.
- ☐ Make appropriate conclusions from observational studies.

2) Every statistical study seeks to answer a research question about a population. Carefully reread the two opening paragraphs above then answer the following questions.

Part A: What question were the researchers seeking to answer? (Make sure your question is directed toward a population.)

Part B: Will the results of the survey yield a valid answer to the research question? Explain your answer.

Part C: If the study gives a valid answer to the research question, give a conclusion. If not, explain the problem (explain how the data are non-representative or biased) and propose a solution.

3) A graduate student gathered 75 local high school student volunteers to estimate the number of hours that U.S. high school students spend studying every week. On average, the students in the sample spent about 12 hours a week studying.

Part A: What question is the researcher seeking to answer? Make sure that the question is directed toward a population.

Part B: Will the results of the survey yield a valid answer to the research question? Explain your answer.

Part C: If the study gives a valid answer to the research question, give a conclusion. If not, explain the problem (explain how the data are non-representative or biased) and propose a solution.

4) A study at the University of Wisconsin in Milwaukee followed teens who smoked marijuana for years. The director of neuropsychology, Krista Lisdahl, found that teens "who began using marijuana [and continued] for many years lost about eight IQ points from childhood to adulthood."[1]

Part A: What question is the researcher seeking to answer? Make sure that the question is directed toward a population.

---

[1] Neighmond, P. (2014, March 3). Marijuana may hurt the developing teen brain. Retrieved November 21, 2014, from http://www.npr.org/blogs/health/2014/02/25/282631913/marijuana-may-hurt-the-developing-teen-brain.

Part B: Can we conclude from this study that many years of marijuana use from childhood to adulthood causes an individual's IQ to decrease? Explain why, or offer an alternate explanation for what was observed.

Part C: What conclusion can be made from this study regarding marijuana use and IQ?

5) What are the characteristics of a well-designed statistical study?

Lesson 15, Part A The Video Game Diet

Consider the following research study.

Research question: Dr. Elizabeth Jackson, a professor at the University of Michigan Health System, surveyed 1,003 middle-schoolers about their snacking habits and their predominant after-school activity (outdoor activities, video games, or watching TV).[1]

Conclusion: Kids who spend most of their time outdoors eat the fewest calories from unhealthy snacks.

(Interestingly, kids who played the most video games ate fewer calories from unhealthy snacks than those who tended to watch more TV.)

Credit: tomispin/Fotolia

1) In a statistical study, a **variable** is an observation that changes from individual to individual.

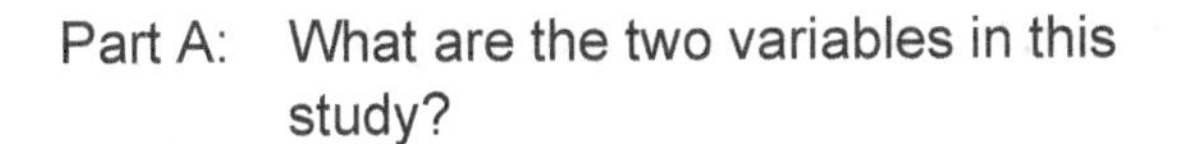

Part A: What are the two variables in this study?

Objectives for the lesson

You will understand that:

- ☐ In experimental studies, researchers manipulate an explanatory variable to observe the effects on a response variable.
- ☐ To manipulate the explanatory variable, researchers divide subjects into similar groups and apply a different treatment to each group.
- ☐ If the groups are truly similar, then differences observed in the response variable might be attributed to the explanatory variable.

You will be able to:

- ☐ Determine when a study design allows a conclusion to be made about cause and effect.
- ☐ Design an experimental study.

---

[1] Singh, M. (2014, March 28). Why playing Minecraft might be more healthful for kids than TV. Retrieved November 23, 2014, from http://www.npr.org/blogs/health/2014/03/28/295692162/why-playing-minecraft-might-be-healthier-for-kids-than-tv.

Part B: Identify the explanatory and response variables in the University of Michigan study and classify them as categorical or quantitative.

Explanatory variable:

Response variable:

Part C: Can researchers in the study conclude that the amount of after-school activity was the cause for the number of calories that kids consumed from unhealthy snacks? Can you offer a possible alternate explanation?

In an **observational study**, researchers simply observe the variables of interest. In an **experimental study**, researchers actively manipulate the explanatory variable.

Part D: Was this an observational or experimental study? Justify your answer.

Part E: In what type of study can we make conclusions about cause and effect?

2) Let's redesign the study about middle-schoolers' after-school activities into an experimental study to see if a cause-and-effect conclusion can be made.

Part A: The explanatory variable is after-school activity. In an experimental study, should the kids (participants) pick an activity according to their preference, or should researchers control this?

Part B: What does it mean in this context to say that "researchers *manipulate the explanatory variable* in an experiment?"

Part C: In an experimental study, the values of the explanatory variable are called *treatments*. What are the treatments in this study?

Part D: Suppose you decide to apply the treatments to different groups by dividing children randomly among three groups. Would there be any reason to think that one group would be very different from another?

Part E: In an experiment, random assignment into groups tends to create groups that are similar. The only real difference from group to group is the explanatory variable that researchers are manipulating. If the children assigned to the video game group consumed fewer calories than the children assigned to the TV group, could there be another explanation other than the activity?

Part F: If the children in the video game group consumed fewer calories than the children in the TV group, what conclusion could be made about the activity?

In experiments, if we apply the various treatments to groups that are similar in composition, then we can conclude that the treatments are responsible for any differences observed in the response variable.

Lesson 15, Part B All Things in Moderation

Chocolate has been a favorite food of humans for over 3,000 years, originating from Central America and Mexico, where it was enjoyed as a drink. It was thought to have certain medicinal properties and was even considered valuable enough to be used as money.[1]

Lately, studies have emerged that also promote the health benefits of chocolate.

1) Have you heard of any health benefits from eating chocolate? If so, what were they?

Credit: Jesse Kraft/123RF

Objectives for the lesson

You will understand that:

- ☐ Conclusions about cause and effect cannot be made in observational studies because of confounding variables.
- ☐ Conclusions about cause and effect can be made in experimental studies where confounding is controlled.
- ☐ Experimental groups created through random assignment can usually control variability among confounding variables.

You will begin to be able to:

- ☐ Analyze a statistical study and identify possible confounding variables.
- ☐ Decide when confounding variables restrict conclusions about cause and effect.
- ☐ Design an experiment that allows conclusions about cause and effect.

In this lesson, we consider how conclusions in statistical studies can be uncertain when researchers do not know why a particular outcome has happened.

---

[1] Benson, A. (2008, March 1.) A brief history of chocolate. Retrieved November 24, 2014, from http://www.smithsonianmag.com/arts-culture/a-brief-history-of-chocolate-21860917/?no-ist.

2) A research study in Sweden asked the research question: "Does chocolate reduce the risk of stroke?"[2]

A sample of 37,000 Swedish men were divided into four groups based on their normal chocolate consumption, ranging from those who ate no chocolate to the group who ate the most (63 grams per week). It is important that the study included so many people because it makes the results more reliable. Participants were followed for 10 years.

The study concluded that men who ate more chocolate had a lower risk of experiencing a stroke.

Part A: Did this study prove that eating chocolate causes a lower risk of stroke? Explain your response.

Part B: Offer an alternate explanation for this lowered risk among chocolate eaters.

3) In Preview Assignment 15.B, you read about **confounding variables**, or factors that give alternate explanations for an effect. When there are confounding variables, researchers cannot make inferences about cause and effect.

Another article[3] pointed out some important facts in the study described. The men who ate the most chocolate tended to be younger, were more educated, and were less likely to smoke or have high blood pressure. They also reported eating more vegetables and drinking more wine. All of these factors have positive health benefits.

Part A: Explain how this information sheds light on the study that says that eating chocolate lowers the risk for stroke? Does it make us more or less certain about the effects of eating chocolate?

---

[2] Larsson, S. C., Virtamo, J., & Wolk, A. (2012). Chocolate consumption and risk of stroke: A prospective cohort of men and meta-analysis. *Neurology*, *79*(12), 1223—1229.

[3] Goodman, B. (2013, January 7). Don't fudge the facts on chocolate studies. Retrieved November 24, 2014, from http://healthjournalism.org/blog/2013/01/dont-fudge-the-facts-on-chocolate-studies.

Part B: Chocolate eaters tended to be more highly educated. What other confounding variables arise from this fact that may also cause a lowered risk of stroke? (What other things do people who are better educated tend to have in common?)

Cause and effect conclusions cannot be made in observational studies because of confounding variables. Experiments allow such conclusions by *holding confounding variables constant* among several groups. To do this, researchers make sure that groups are similar in terms of education, age, eating habits, and so on. When groups are similar, confounding variables are controlled (not varying), and it becomes easier to argue that a treatment (like chocolate) is the cause of an observed effect (like fewer people having a stroke).

To control for the confounding variables, researchers divide participants into similar groups. Some groups are **treatment groups**—they receive a real treatment. Other groups are **control groups**—they receive a **placebo** (a fake treatment) or nothing at all. Control groups are used as a basis for comparison to the treatment groups.

4) We want to design a new study to infer whether eating chocolate causes lowered risk of stroke. To do this, we compare a large group of chocolate eaters with another large group of non-chocolate eaters.

   Part A: Which is the control group? Which is the treatment group?

   Part B: Would it make sense to allow the study participants to choose which group they are in? Explain the possible consequences of such an approach.

   Part C: As researchers, it is our job to create control and treatment groups so that confounding variables, such as education or income, do not interfere with the outcome. What should be our goal as we decide how to assign subjects to groups?

In **controlled statistical experiments**, researchers try to minimize the effects of confounding variables by creating treatment and control groups that are as similar as possible. Groups are similar when confounding variables are held constant (on average) from group to group.

5) You have gathered a group of volunteers who are all in their early 20s. They come from various socioeconomic backgrounds, and none eat chocolate, but they are all willing to try it for the sake of the experiment. You want to divide them into two groups (control and treatment), and you will monitor their health for four decades.

Part A: Devise a method for dividing the volunteers into groups that will minimize confounding in the long-term results. Be specific.

Part B: What are the explanatory and response variables in this study? Are they quantitative or categorical?

Part C: With your method of dividing the volunteers into groups, suppose that the decades-long study concludes with a significantly lower rate of stroke among those who ate chocolate. Write a sentence that infers a conclusion for the study.

Lesson 15, Part C The Power of the Pill

In a study conducted with 66 chronic migraine headache sufferers, patients were given a pill to treat their pain. Pain levels were measured before and after taking the pill. On average, pain was reduced by 26% after 2½ hours.[1]

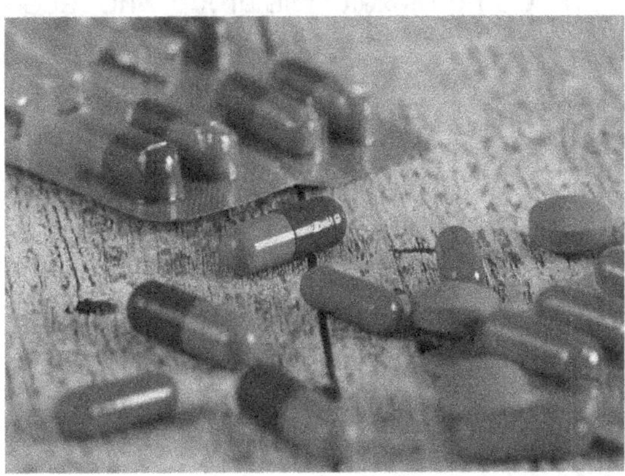

Credit: SakisPagonas/Fotolia

1) Did this study prove that the treatment is effective in reducing pain? Think about whether there are confounding variables that offer an alternate explanation for the patients' improvement.

Objectives for the lesson

You will begin to understand that:

- ☐ The belief that a treatment will help may bring about improvement for some ailments. This improvement is known as the placebo effect.
- ☐ The belief that a treatment may bring about improvement is a confounding variable unless members of a control group receive a placebo.
- ☐ **Blinding** is when patients do not know if they are receiving a treatment or placebo; **double blind** is when neither the patients nor their research contact knows who is receiving a treatment or placebo.

You will begin to be able to:

- ☐ Decide when an experiment should introduce a placebo to control confounding.
- ☐ Design a double-blind study that uses a placebo to control confounding.

The treatment that was given to the 66 patients in the migraine study was a **placebo**. It was a pill filled with inactive ingredients. Nonetheless, pain levels decreased by 26%. This positive response to a fake treatment is known as the **placebo effect**.

---

[1] Knox, R. (2014, January 10). Half of a drug's power comes from thinking it will work. Retrieved November 24, 2014, from http://www.npr.org/blogs/health/2014/01/10/261406721/half-a-drugs-power-comes-from-thinking-it-will-work.

2) In a different study, patients were divided randomly into two groups. One group received a well-known migraine headache medication called Maxalt, while another group received nothing. On average, there was a 40% decrease in pain levels after 2 hours for the group receiving Maxalt. The group who received nothing had a 15% decrease in pain.

Part A: The patients were divided randomly, so the groups should be roughly similar. Unfortunately, there is still a confounding variable in this study. What is the confounding variable?

Part B: Can we say that Maxalt causes a decrease in pain levels for migraine sufferers based on the results of this study? Explain your answer.

Part C: When there is a confounding variable in a study, there is an alternate explanation for the changes observed. What is the alternate explanation for the changes in this study?

Part D: What is a possible remedy for the confounding caused by providing the Maxalt pill?

Part E: If we give a placebo to one group and Maxalt to the other, which group is the treatment group, and which is the control group?

Treatment group:

Control group:

Part F: If we give a placebo to one group and Maxalt to the other, and the patients know what they are getting, do we still have a confounding problem? Explain your answer.

When we say that an experimental study implements **blinding**, we mean that patients do not know that they are receiving a placebo. Usually, the volunteers sign a statement that they understand they may receive an active treatment or they may receive a placebo.

3) Some studies have shown that the person who gives the placebo to the patient can influence its effect. For instance, if the doctor or nurse says that it is a very powerful medication that will help them, the placebo effect can be stronger.

   Part A: Should the person who gives a placebo to a patient know that it is a placebo? Explain your answer.

   Part B: In a **double-blind** study, neither the patients nor those who interact with them know who is receiving a placebo. If an experiment is double blind with similar groups created by random assignment, with one group receiving a treatment and the other receiving a placebo, and the treatment group improves significantly more than the control group, can we conclude that the treatment is effective?

Depending on the ailment, the placebo effect can be almost as (or more) powerful as the drug itself. In the study described at the beginning of this lesson, patients were given pills with different information. When patients were given a placebo but told it was Maxalt, the pain reduction was statistically similar to when patients were given Maxalt, but told it was a placebo.

The most powerful combination was when patients were given Maxalt and told it was Maxalt.

4) Suppose that, in an experiment, three groups are given three different dosage levels of a new experimental medication: 10mg, 20mg, and 40mg. If each group experienced some improvement in the way that was intended, why had this improvement occurred?

   a) The medication was effective.
   b) The experiment was not effective.
   c) Without a control group, we cannot be sure that the improvement was due to the medication.

Lesson 15, Part D Designing an Experiment

About 90% of adults worldwide are infected with the virus that causes cold sores.

You work for an employer who has developed a new cold sore cream that needs to be approved by the U.S. Food and Drug Administration before it can be sold. You are assigned to a team that will design and implement a study to determine the effectiveness of the new cream.

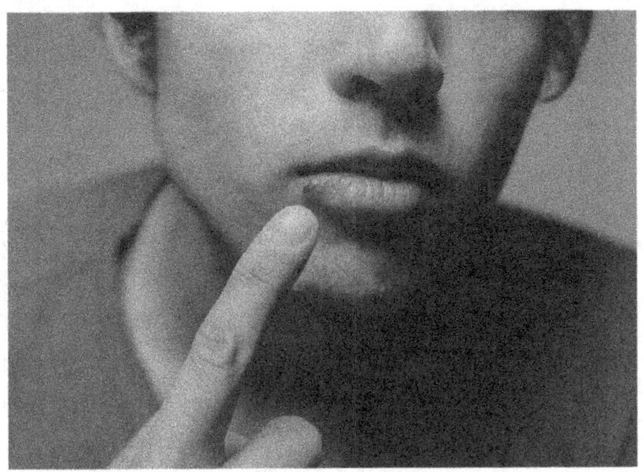
Credit: LoloStock/Fotolia

1) Is the goal of this study to make an inference about a population or a treatment?

Objectives for the lesson

You will understand that:

- ☐ Double-blind experiments are the best way to conduct medical experiments with human subjects.
- ☐ When conducted properly, a double-blind medical experiment allows researchers to infer a causal relationship between a treatment and its response.
- ☐ When **blocking** is implemented in an experiment, it helps minimize variability among groups so that the response to a treatment is more obvious.

You will be able to:

- ☐ Design a double-blind experiment with blocking.
- ☐ Make a conclusion that is appropriate to the results of an experimental study.

2) Your team has gathered 76 volunteers for the study. There are 38 women and 38 men; each has the virus that causes cold sores.

   Part A: Is it a problem that the participants are volunteers? Explain your answer.

Observational studies use representative samples to describe larger populations. This study is an experiment that describes the effects of a treatment. It is best when experiments have representative samples so that results can be applied to entire populations. This is not always possible, but meaningful inferences can be still be made with the understanding that they may not apply to all population members.

Part B: You have some concerns that men and women may react differently to the medication. Work with your group to devise a plan that will deal with this difference.

If we feel that there are groups (or **blocks**) of individuals in a study that may react differently to a treatment, such as men and women or possibly age groups, one strategy to deal with these differences is to conduct the experiment on each group separately. This method is called **blocking**. Each block contains members that have something in common, like gender or age. Each block is then divided into treatment and control groups by random assignment. Blocking helps cut down on variability within control and treatment groups so that treatment effects are easier to identify.

3) Your group has decided to minimize variability within groups by creating two blocks—one with the female volunteers and the other with the male volunteers. Each block has 38 members and needs to be divided into control and treatment groups.

Part A: What is our goal in terms of the differences between the groups?

Part B: What process should be applied to divide members into control and treatment groups?

4) We now have two control groups (one female and the other male) and two treatment groups (one female and the other male). The two female groups are similar; the two male groups are also similar.

Part A: What will you give the treatment groups?

Part B: What will you give to the control groups?

Part C: Will you tell the members of the groups exactly what they are getting? Explain your reasoning behind this decision.

You have decided to measure the effectiveness of the treatment based on the number of days until the cold sore is gone. You instruct each study participant to begin using the cream at the first sign of their next cold sore. They are to apply the cream 4 times a day until the sore is gone.

5) The data for the female control and treatment groups are provided below. These values represent the number of days for the sore to disappear.

Control: 22, 19, 14, 22, 21, 16, 5, 20, 11, 16, 14, 15, 17, 13, 16, 10, 17, 16, 11

Treatment: 11, 17, 8, 16, 15, 12, 11, 12, 12, 9, 11, 19, 16, 13, 6, 15, 12, 9, 10

Part A: Make a dotplot for each data set. Be sure that you scale both dotplots the same size.

Part B: Did everyone in the treatment group heal more quickly than everyone in the control group? Explain your answer.

Part C: Calculate the average time for the female control group, rounding to the hundredths place. Plot that value on your dotplot as an "X" on the horizontal axis.

Mean of female control group = _____

Part D: Calculate the average time for the female treatment group, rounding to the hundredths place. Plot that value on your dotplot as an "X" on the horizontal axis.

Mean of female treatment group = _____

Part E: Make a conclusion that is appropriate for this type of study.

Lesson 15, Part E In Conclusion

In a recent study on helping obese people with diabetes, a group with Type 2 Diabetes was divided randomly into three groups.[1] A third of them received *gastric bypass surgery* (in which most of the stomach is surgically bypassed), and another third received another surgery called *sleeve gastrectomy* (where most of the stomach is surgically removed). The remaining participants were treated with medication alone.

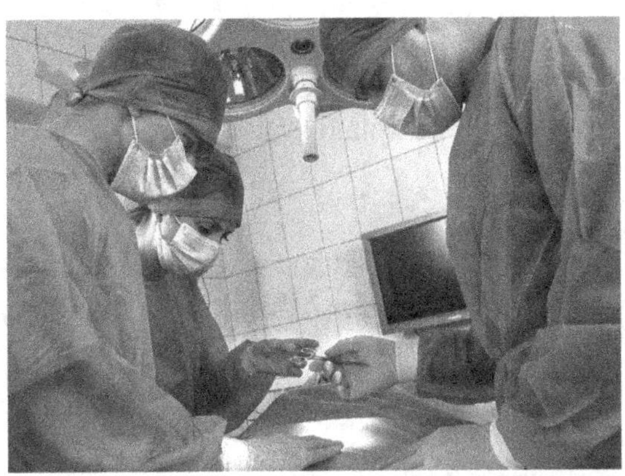

Credit: BillionPhotos.com/Fotolia

1) Is the study observational or experimental?

Objectives for the lesson

You will understand that:

- ☐ In observational studies, blocking can be implemented to help control confounding variables. This strategy can make associations more certain. The conclusions should not imply cause.
- ☐ In experimental studies, blocking is implemented to makes the study more sensitive to change from treatments. Causation can be inferred.

You will be able to:

- ☐ Make appropriate conclusions from observational studies and from experimental studies.
- ☐ Identify problems in studies that prevent researchers from making inferences to populations or treatments.

2) More than 33% of those who had gastric bypass surgery met blood sugar targets after 3 years, compared to 24% of those who had the sleeve gastrectomy, and just 5% of those with medication alone. For either surgery, the proportion of those who met blood sugar targets was significantly higher than for those on medication alone.

---

[1] Shute, N. (2014, March 31.) Weight-loss surgery can reverse diabetes, but cure is elusive. Retrieved November 24, 2014, from http://www.npr.org/blogs/health/2014/03/31/297160022/weight-loss-surgery-can-reverse-diabetes-but-cure-is-elusive.

Part A: Would it make sense to implement a control group with a placebo and blinding for this study?

Part B: Is it possible that the placebo effect was the cause for the results observed? Explain your answer.

Part C: Would it make sense to implement blocking to minimize variability within treatment groups? What blocks would you suggest?

This study did not have a typical control group. If patients are very ill, we don't want to not give them any treatment at all. In such a case, the "control group" is a group that gets the traditional treatment, such as the group in this study who received medication rather than one of the surgeries.

Part D: Taking any weaknesses of this study into account (no real control group), write a conclusion that addresses the effectiveness of each treatment in the study.

3) A recent study followed a sample of 2,700 men and women from the U.S. for 25 years. [2] This research was a part of the "Coronary Artery Risk Development in Young Adults" study. One of the conclusions from the study was those participants who exercised as teens and young adults performed better on memory and problem solving tests in middle age. These results occurred even when blocking was implemented to account for unhealthful factors such as smoking, diabetes, and high cholesterol.

The sample included people of several (not *all*) ethnic backgrounds from the United States. These groups and their subcategories (gender and age) were selected in proportions matching the population.

Part A: Was this study observational or experimental?

---

[2] Singh, M. (2014, April 2.) Run when you're 25 for a sharper brain when you're 45. Retrieved November 24, 2014, from http://www.npr.org/blogs/health/2014/04/02/297910425/run-when-youre-25-for-a-sharper-brain-when-youre-45.

Part B: Can we use the results of this study to make an inference about all people who live in the U.S.? If not, then to whom can the results be generalized?

Part C: What does it mean that blocking was implemented to account for unhealthful factors such as smoking, diabetes, and high cholesterol?

Part D: Can we conclude that exercising as teens and young adults caused better problem solving and memory skills in middle age? If the relationship is not causal, can you offer an alternate possible explanation?

Lesson 16, Part A Education Pays

The graph to the right shows data from a study conducted by the United States Bureau of Labor Statistics.[1]

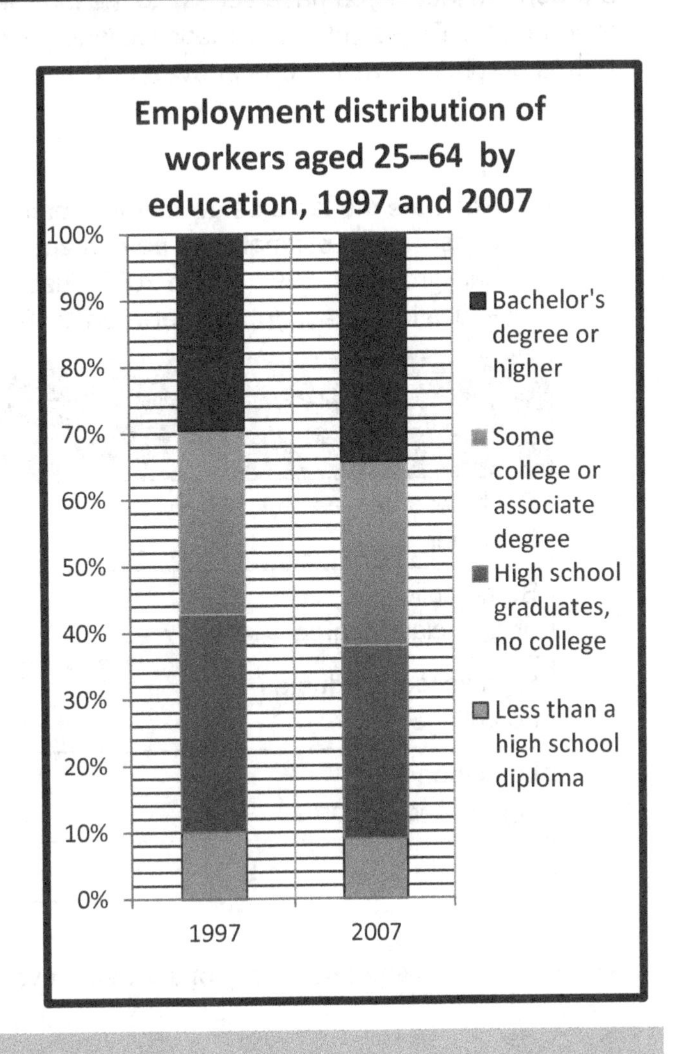

1) Let's figure out the story that this graph is telling. A good first step is to break down the graph into different pieces.

Part A: Identify and state what each axis measures.

Part B: What do the different colors in the graph represent?

Part C: What population is represented in the graph? All persons or all workers?

Objectives for the lesson

You will begin to understand that:

- ☐ A stacked column chart can display relative frequency data for different categories or over different time periods.

You will begin to be able to:

- ☐ Determine information from a stacked column graph.
- ☐ Analyze data in a stacked column graph and write a brief summary of the information.

[1] Source: U.S. Bureau of Labor Statistics. (2008, July). *Older workers*. Retrieved December 10, 2014, from http://www.bls.gov/spotlight/2008/older_workers.

2) Sometimes the media does a good job of including all the information necessary to analyze a graph. (In fact, if you have access to the internet, you can find the actual data for this report on the Bureau of Labor Statistics website.) However, you often have to estimate data from a graph such as the one above.

Part A: Recreate the table below in your notes. Complete the table by using the graph in question 1 to estimate the percentage of workers with each educational background in 1997 and in 2007. Also determine the absolute change in each educational background between 1997 and 2007.

Level of Education	1997	2007	Absolute Change in Percentages
Bachelor's degree or higher			
Some college or associate degree			
High school graduates, no college			
Less than a high school diploma			
Total			X

Part B: Write a short summary of the data in your table.

Part C: What level of education was most common of workers aged 25–64 in 1997?

Part D: What level of education was most common of workers aged 25–64 in 2007?

3) Let's look at the data in another way.

Part A: Add a new column to the right of the table, containing the relative change in percentages for each education category.

Part B: Identify the category that showed the largest change between 1997 and 2007. What was the change?

Part C: What percentage of the workforce in 1997 had no education at the college level? How did that compare to the percentage of the workforce in 2007 with no college-level education?

4) Analyze the graph and the table of values, and write a summary about how the educational composition of the workforce changed between 1997 and 2007.

5) Did you find the stacked column graph easy to analyze? Create another graphical display that illustrates the same information.

Lesson 16, Part B Looking for Links

In Preview Assignment 16.B, you read about the observational study from the Centers for Disease Control and Prevention.[1] This study focused on the intelligence quotient (IQ) of children aged 8 years who were identified with autism spectrum disorder (ASD).

Credit: ladyalex/Fotolia

1) By looking at the graph on the next page, what observations do you have, or what patterns do you see?

Objectives for the lesson

You will understand:

- ☐ How stacked column graphical displays of data can be useful to show correlation between different variables.
- ☐ That correlation between variables is not the same as causation.

You will be able to:

- ☐ Analyze data given in a stacked column graph and write a brief summary of the information.

[1] Centers for Disease Control and Prevention. (2014, March 28). Prevalence of autism spectrum disorder among children aged 8 years: Autism and developmental disabilities monitoring network, 11 sites, United States, 2010. *Morbidity and Mortality Weekly Report, Surveillance Summaries, 6*(2). Retrieved January 4, 2015, from http://www.cdc.gov/mmwr/preview/mmwrhtml/ss6302a1.htm#Fig2.

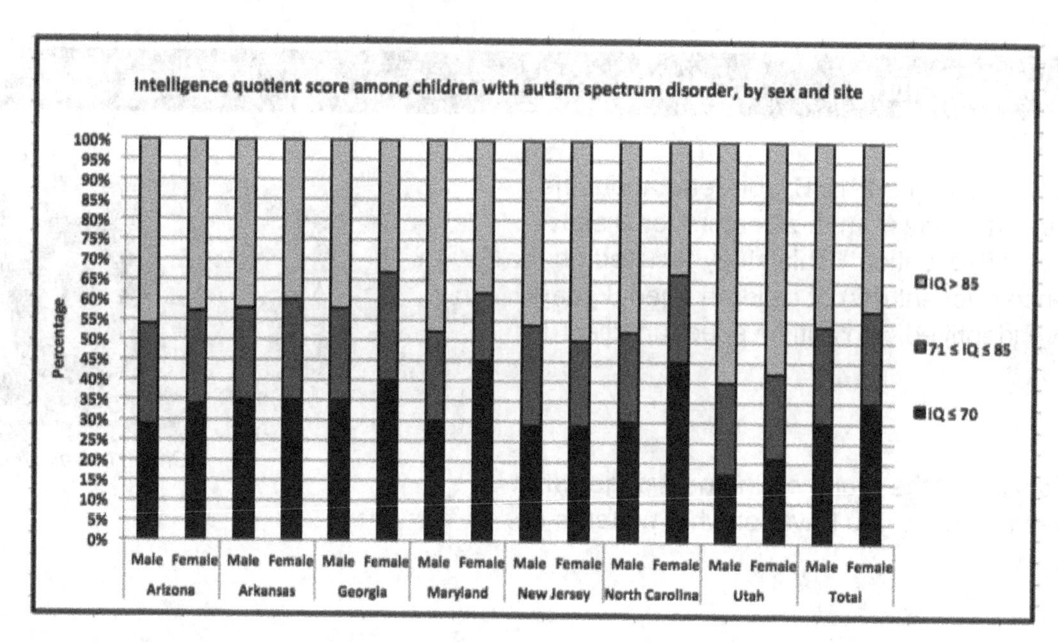

2) Revisit the graph and your estimates from the Preview Assignment. Recall that a borderline IQ is defined as the 71-to-85 range (the middle segment of each bar).

Part A: Compare the percentage of males with ASD in Arizona who have borderline IQ with the percentage of males who have average or above average IQ. Make a similar comparison for females with ASD in Arizona.

Part B: Find the state (or states) with the largest percentage of males with ASD who have average or above average IQ, and estimate that percentage. Do the same for females.

Part C: Which state has the highest percentages of males and females with ASD who have intellectual disabilities (bottom segment)? Which state has the lowest percentages of males and females with ASD who have intellectual disabilities?

3) Explain why each statement is <u>not</u> supported by the graph. If possible, correct each statement.

Part A: In Arizona, about 32% of boys with intellectual disabilities were diagnosed with autism spectrum disorder.

Part B: In Arizona, about 32% of boys have intellectual disabilities.

Part C: In Arizona, girls are not as smart as boys.

Part D: Autism causes low intelligence.

4) What can you say about the IQ levels of females with ASD compared with IQ levels of males with ASD for all of the states that were in this study? (Hint: Look at the "Total" columns.)

5) One area of study is the association between intellectual ability and diagnosis of autism spectrum disorder. Does the graph provide information about this relationship?

6) Autism Speaks, an autism advocacy organization in the United States, estimates that boys are nearly five times more likely than girls to be autistic. How does this graph relate to the organization's claim?

Lesson 16, Part C It's About Time!

In Lessons 1, 8 and 16, you submitted data about the number of hours you spend working on math homework outside of class alone and with others.

1) Think about the amount of time you spend outside of this class each week working on course material. Include tutoring, group work, time you spend studying alone, and any other work related to class.

Credit: Photographee.eu/Fotolia

Part A: Has your study time changed during the term?

Part B: Make a hypothesis: Do you think people in this class are studying more or less outside of class as the term has gone on?

Part C: Do you think there is a relationship between time spent outside of class studying and the course grade?

Objectives for the lesson

You will understand:

- ☐ The usefulness of grouping data to look for patterns.
- ☐ The impact of your study time on success in a mathematics course.

You will be able to:

- ☐ Use a spreadsheet to sort specific information for analysis and graphing.
- ☐ Use a spreadsheet to build a stacked column graph.
- ☐ Analyze data and the related stacked column graph and make conclusions about the pattern of the data.

In this lesson, we will analyze our class data to understand how time spent studying for this class has changed over the term. This data set contains information about hours spent on this class—outside of class alone and with others—from the first week (Lesson 1), and Lessons 8 and 16. We will group, or aggregate, the data and make a stacked column graph showing the data over these three time periods.

2) Open the spreadsheet of your class members' study time. Inspect the data, ensuring that you understand what you see in the columns.

Part A: Do you see any patterns at this point?

Part B: Sort the data in each total column in increasing (ascending) order. Highlight the data in the column titled "Total hours studied per week" for Lesson 1, and then go to the Data menu and select "Sort." Remember, you can always click on the "?" help button if needed.

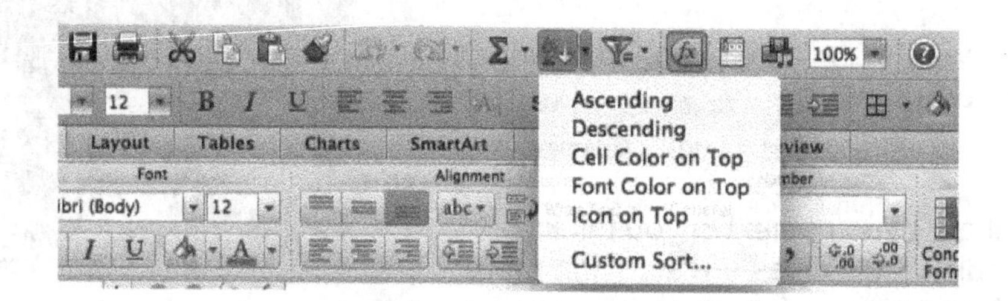

You may see a dialogue box asking whether you want to expand the selection. Click the button that says "Continue with the current selection" and in the next window, click the small box that says "My list has headers." Then select the appropriate column from the dropdown menu and click "OK." Check that the data has sorted appropriately. Repeat these steps to sort the columns representing total study time for Lessons 8 and 16.

Part C: Make a table that aggregates the data of total number of hours spent studying outside of class (alone and with others) into three categories:

- Students who spend 1 hour or less outside of class for every hour they spend in class.
- Students who spend more than 1 hour but less than 3 hours outside of class for every hour they spend in class.
- Students who spend 3 or more hours outside of class for every hour they spend in class.

For each week, use the sorted data in the corresponding column and count the frequencies, or the number of students in each category. Make a table in your spreadsheet to record your data. For example, for a course that meets for 3 hours per week, the table should look similar to the one shown below.

Hours Spent Studying Outside of Class	Student Data Lesson 1	Percent Lesson 1	Student Data Lesson 8	Percent Lesson 8	Student Data Lesson 16	Percent Lesson 16
< 3 hours						
3–9 hours						
> 9 hours						
Total number of students						

Part D: Calculate the percentage of students in each category.

Part E: Do you see any patterns from these tables?

Part F: Now use the spreadsheet Chart feature to create the stacked column chart. Highlight the first, third, fifth and seventh columns of the data in the table, go to "Charts" and click on "100% Stacked Columns." Use the Chart Layout tab to label the axes and give the chart a title.

Part G: Use the information in the chart to write a description of how the study time of students in the class changed as the term went on. Use quantitative information in your statements. Does the information in the chart match your hypothesis about student work?

3) How does your class data compare with the CCSSE survey data presented in question 1 of Preview Assignment 16.C?

4) Answer the following questions to discuss the advantages and disadvantages of a stacked column graph for the class data.

Part A: Is a stacked column graph a good tool to display this data? Why or why not?

Part B: What other graphical displays could be used for this data?

Lesson 16, Part D Connecting the Dots

We live in a world that creates and uses a lot of data and information. This information is used in our society for different purposes such as marketing and politics. Data visualization is the study of different ways to display data through visual representations, such as graphs. The purpose of data visualization is to use appropriate tools, such as graphs, to help people understand data.

In this lesson, we will look at some complicated graphs that come to life with technology. We will also see how these graphs incorporate more variables than the graphs that we have already seen.

Credit: polygraphus/Fotolia

1) Think about advertisements you see throughout your day—for example, on the bus or the highway. What catches your eye or draws your interest? List one example of a graphical display of data in a commercial or advertisement.

Objectives for the lesson

You will understand that:

- ☐ Visualizing data in different types of graphical displays provides tools that can help people understand patterns in the data.
- ☐ Data can help identify and/or understand historical events and how the world has changed over time.

You will be able to:

- ☐ Identify different variables in a graph with several variables.
- ☐ Analyze motion bubble charts to identify trends and patterns.

In Preview Assignment 16.D, you were introduced to some interactive graphs from the Gapminder website. These graphs are dynamic and show how variables change over time. Many of the graphs are *bubble charts*, which use bubbles to represent information. The area of each bubble corresponds to data from a new variable. These dynamic bubble charts can present information using more than two variables, conveying much more information than the other charts and graphs that we have seen so far.

Note that some of the graphs include data from over 200 years ago. You may wonder how that data were collected. For the most part, no such data exist. Instead, these data were estimated or determined using statistical models. Life expectancy data, for example, were modeled using some basic assumptions. One assumption is that most countries in the 1800s had very low levels of life expectancy, typically between 25 and 40 years. The other assumption is that once a country goes through a period of gains in health, then life expectancy should improve. These models were determined by studying historical and demographic accounts. Gapminder emphasizes that the historical data are very rough estimates.

2) Let's go back in time to 1800. Think for a minute about what the world was like then, so that you will have a context for the information that you will see.

   Part A: What was the economy mainly based on at that time?

   Part B: What was health and medical care like at that time? How does that relate to life expectancy?

3) Return to the graph you explored in the Preview Assignment, found at www.gapminder.org/world/. Slide the time back to 1800 and notice that the graph updates itself.

   Part A: What can you say about the countries in the world at that time? What was the richest country in 1800? How does this picture compare to the picture in 2012?

   Part B: Click "Play" and watch the graph go into motion. What patterns do you see? What do you think of the information that is being presented? (Hint: You will need to "zoom in" to the lower left part of the graph.)

   Part C: How have the historical events over the past 200 years impacted life expectancy or the economy in certain countries? Use the graph to justify your conclusions.

4) In the next set of questions, you may want to make sure the "Trails" box is checked so that you will be able to see clearly a country's position changing over time.

Part A: In the box to the right, select Germany, Italy, Russia, United Kingdom and United States. Move the time bar ahead to 1910, then click "Play" and stop at 1920. What do you see?

Part B: Click "Play" again and stop at 1933. What happened with Russia's data over this time frame?

Part C: Deselect all countries, then select Japan and the United States. Set the year at 1935 and click "Play." What do you observe?

5) Think back to what you have seen on this graph and in Preview Assignment 16.D. In general, if a country gets richer, will it see better life expectancy?

If you are interested in learning how to construct these types of graphs, you may want to investigate Google's Public Data Explorer, https://www.google.com/publicdata/directory.

Lesson 16, Part E Big Data

We now move from data that you can organize, analyze and present easily to **big data**. The term says it all. Big data is different because it is so large and complex, structured and unstructured, that using it has required new developments in software and hardware.

1) Recall from Preview Assignment 16.E that the three characteristics of big data are volume, velocity, and variety.

Credit: Mikko Lemola/Fotolia

Part A: Name a company that acquires a large amount of data.

Part B: What kind of data do they acquire?

Part C: How fast does the data arrive?

Part D: What can they do with the data to improve what they do?

Objectives for the lesson

You will begin to understand:

- ☐ The role that **big data** plays in your life.
- ☐ What an attribute table is.
- ☐ The use of logical statements in mining big data.

You will be able to:

- ☐ Convert from degrees, minutes, and seconds to a decimal equivalent.
- ☐ Choose the appropriate logic and write a query to identify a subset of a population.

2) One of the tools used to organize and analyze big data is GIS software. A geographic information system (GIS) allows us to visualize, question, analyze, interpret, and understand data so that the data reveal relationships, patterns, and trends. A GIS can help us to recognize and analyze the spatial relationships that exist within digitally stored spatial data.

Mapping spatial data requires that there be a coordinate system by which we can define the location of any point on the Earth's surface.

Part A: How is a coordinate system of the Earth's surface different from the x- and y-coordinate system with which you are most familiar?

Part B: Longitude and latitude are the coordinates used in mapping spatial data on the Earth's surface. The coordinates of the Alamo are roughly 29° 25' 32.5338" latitude and –98° 29' 9.9378" longitude. Express the coordinates of the Alamo as an ordered pair in decimal form, rounded to the nearest thousandths.

Part C: The Texas State Capitol Building on 1100 Congress Avenue in Austin is located at (30.275, –97.740). How can you tell from the coordinates that the Capitol in Austin is north and east of the Alamo in San Antonio?

3) Big data can be a collection that is spatially related; that is, the data can be represented by locations on a map. That data usually has additional attributes other than location. The following figure shows a map with three layers: a major-street map layer of a county; a hydrology layer showing major water features (light shading); and a student layer showing a hollow, green (dark) dot for every one of the 149,621 K–12 students registered in the school district in that county during the 2012–13 school year.

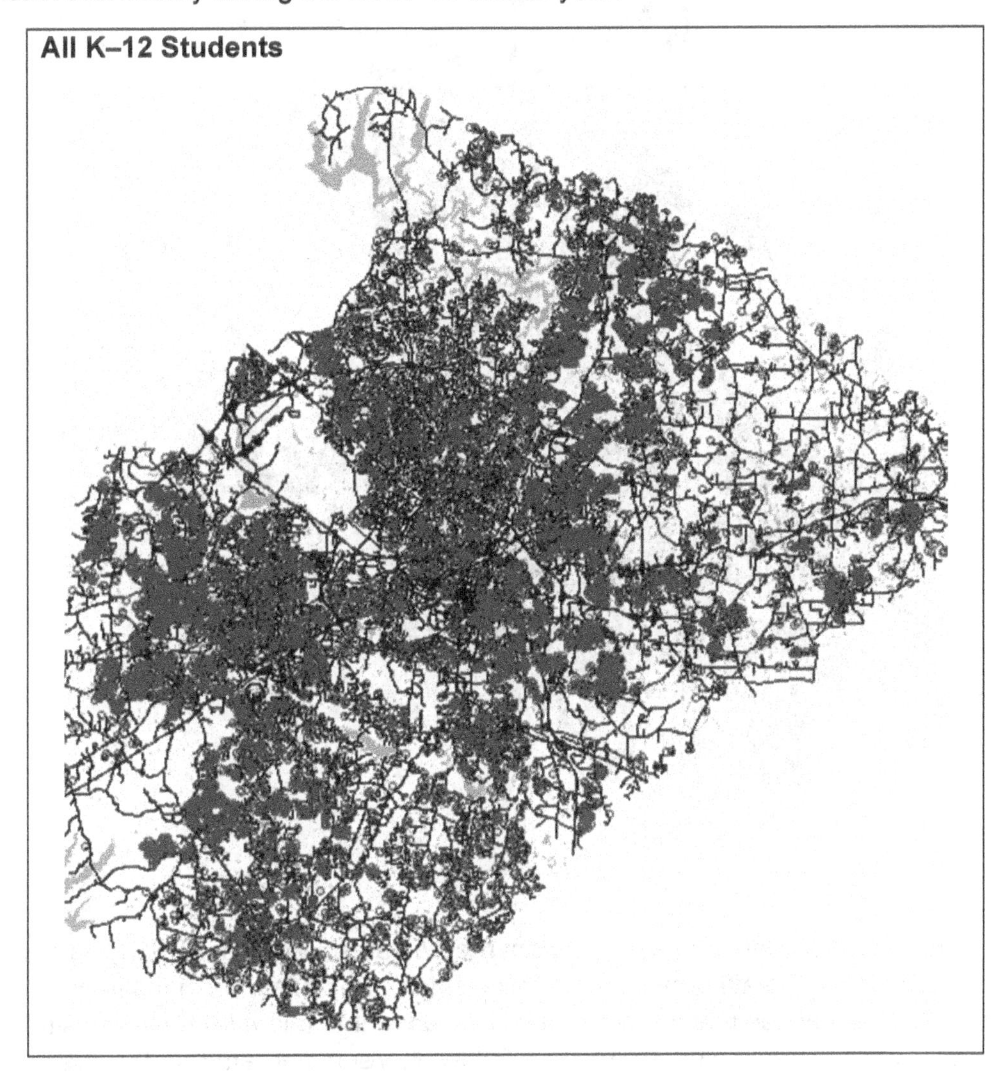

Part A: Some areas appear to be colored solid green even though the students are marked by hollow dots. Explain why that is the case.

Part B: Each dot is placed on the map using geocoding tools that use information from the student file. In many cases, the student's address is used to place the student on the map. What other information do you think might be in the student file?

The following figure shows the location of every 8th-grade student by a hollow, red dot. Eighth-graders comprise a subset of all K–12 students.

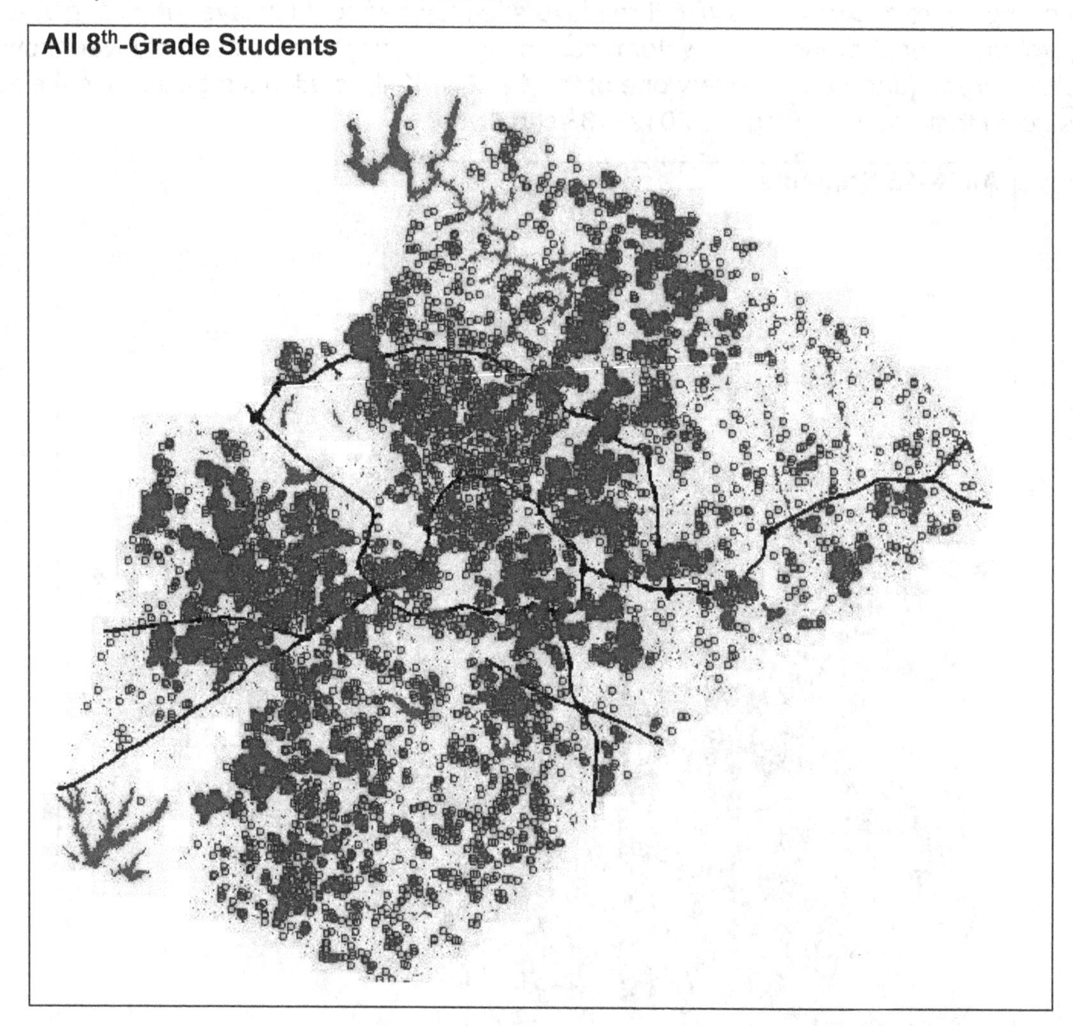

Each dot represents a student and each student record has several accompanying attributes, represented by fields in the attribute table as follows (left). Each field provides information about that student. You cannot see many fields that are included that might relate to the student's academic record or socioeconomic status. The following Venn diagram (right) represents the relationship between the two large graphs.

K-12 Students
Attribute Table

ID	Last_Name	First_Name	Curr_School	Curr_Grade	Sex
112358	Gauss	Johann	1777	8	M
110110	Poisson	Denis	1781	8	M
141421	Canton	Georg	1845	8	M

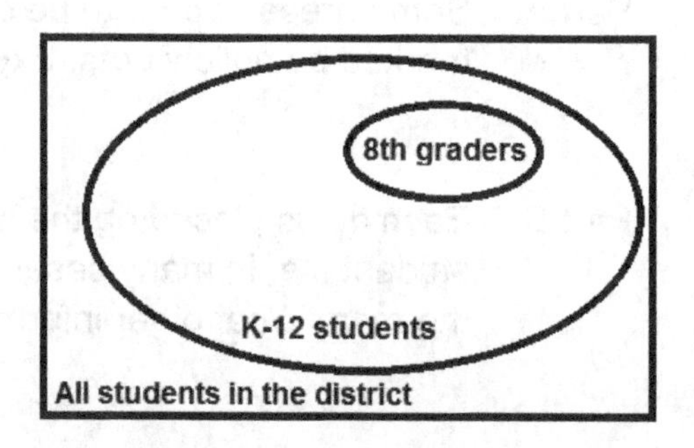

The following figure shows the location of every 8th-grade, male student. These maps are generated by logical queries applied to the entire set of K–12 students.

The school system will need to “mine the data,” locating (or listing), for example, all of the students who have free or reduced lunch. As another example, the system may need to locate all of the middle school students who scored less than 50% on an academic measure. Software allows this large amount of data to be mined easily using queries.

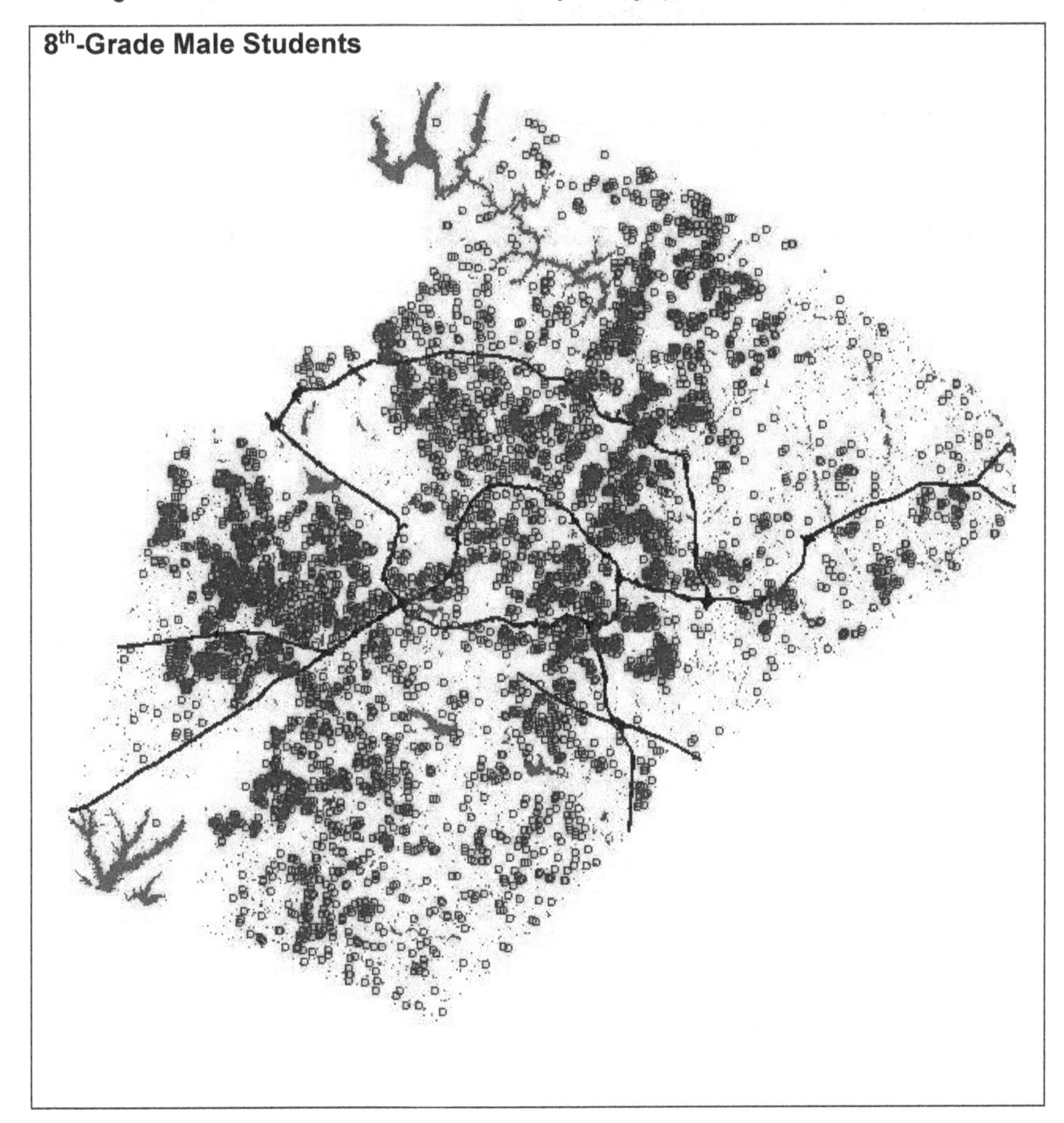

Terms are statements that can be true or false. They often use mathematical operators such as “=” (equal to), “<” (less than), “>” (greater than), or “<>” (not equal to). Logical operators, **AND**, **OR** and **NOT**, join terms to form a query. A data row is placed in the output if the conditions of the query are satisfied.

Part C: Which of the following queries would generate the list of 8th-grade, male students? Indicate "yes" or "no."

i. SEX = M AND CURR_GRADE = 8

ii. SEX = M OR CURR_GRADE = 8

iii. CURR_GRADE < 9 AND SEX = M

iv. CURR_GRADE < 9 AND CURR_GRADE > 7 AND SEX = M

v. CURR_GRADE = 8 AND SEX <> F

Part D: Which of the following figures are most appropriate if the shaded area is to represent 8th-grade (E) male students (M).

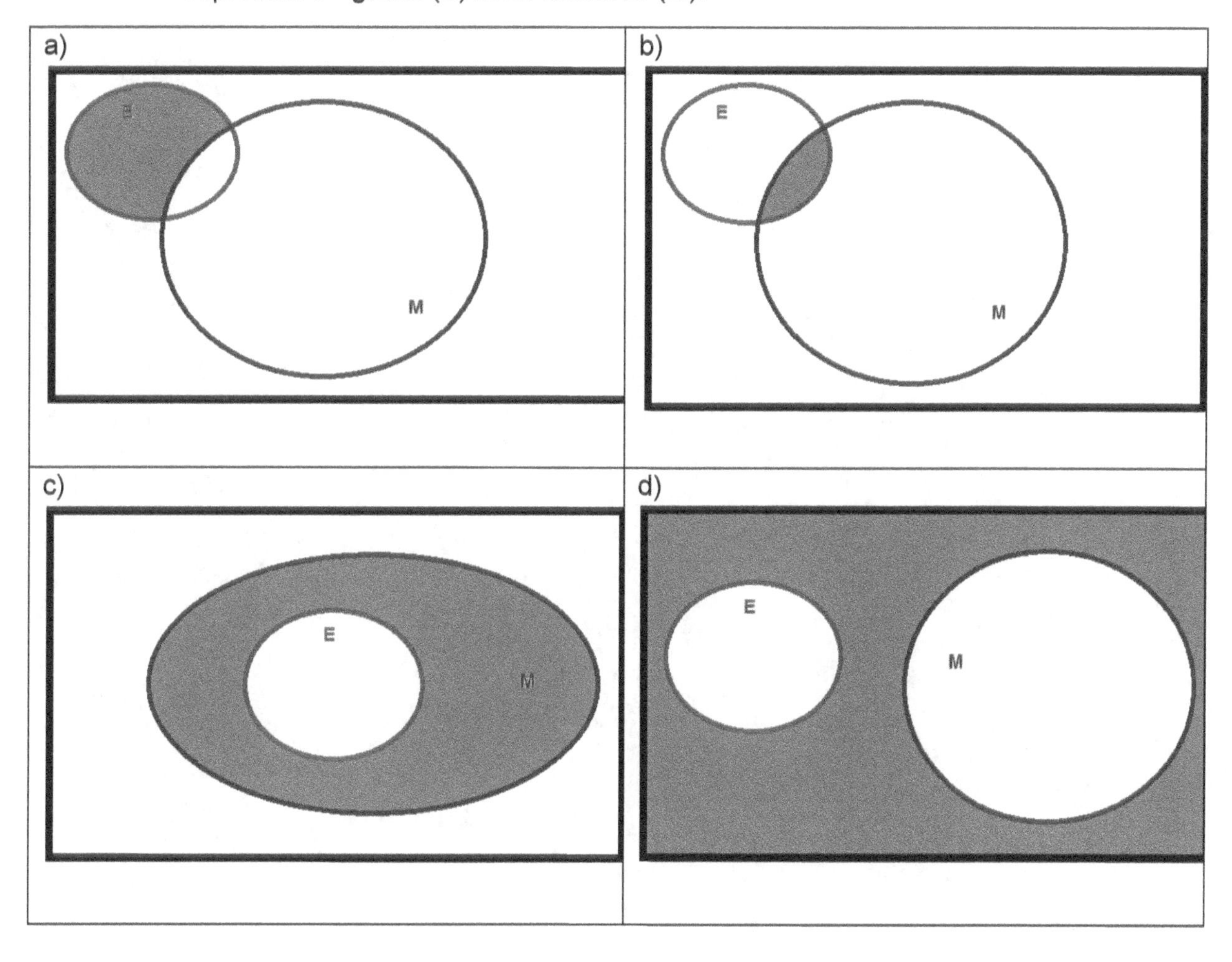

Lesson 16, Part F Big Brother – They're Watching!

From trendy boutiques to big box stores, technology is creeping into store aisles and keeping track of where shoppers walk and what they touch. You watched a video about this retail research in Preview Assignment 16.F.

1) The **heat map** at the right shows the layout of a store. The areas that are shaded black are displays. The rest of the store is color-coded by the amount of foot traffic in that area. The areas shaded red are high-density areas; the areas with little shading are low-density areas.[1]

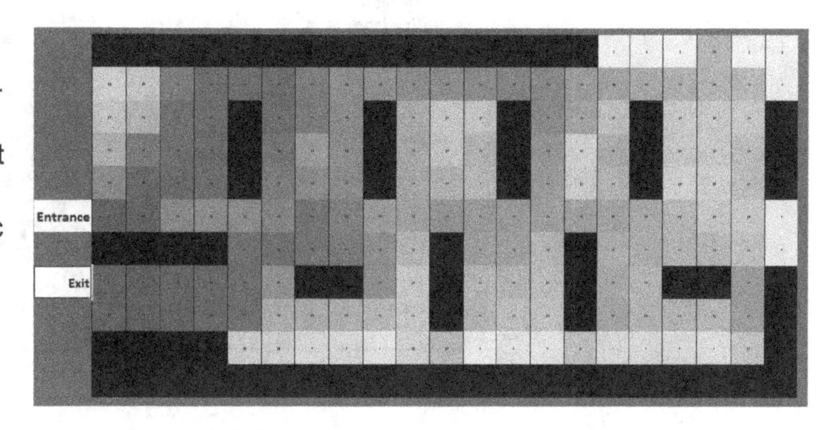

Part A: According to the video, how are retailers collecting this information?

Part B: How could the retailers use this information?

Part C: What is the unit of measurement for the shaded cells?

Objectives for the lesson

You will understand that:

- ☐ A heat map is a color-coded display of a two-dimensional matrix.

You will be able to:

- ☐ Analyze and draw appropriate conclusions from heat maps.
- ☐ Answer questions using density information from a heat map.

---

[1] Dwoskin, E., & Bensinger, G. (2013, December 9). Tracking technology sheds light on shopper habits: Mall operators, retailers monitor patterns and actions. *The Wall Street Journal.* Retrieved December 28, 2014, from http://online.wsj.com/news/articles/SB10001424052702303332904579230401030827722.

2) Banks often have multiple branches in a single city. The main office of one such branch regularly receives customer reviews (where 1 = Poor and 5 = Excellent) from each of the three departments of each branch office. One way to analyze this data would be to display the average of all of the ratings for each branch and for its individual departments in a heat map, such as the one shown below.

	A	B	C	D
1	Average Rating from 1,000 Evaluations	Dept A	Dept B	Dept C
2	Location 1	3.80	3.90	3.70
3	Location 2	3.83	3.73	3.55
4	Location 3	4.12	3.57	3.80

The colors help to identify favorable ratings (green) from less favorable ratings (red).

The nine cells are color-coded based on the average rating. While data can be cumbersome to analyze, the heat map quickly shows whether there are departments that need to improve. This can be accomplished using the Conditional Formatting feature of a spreadsheet program.

Example 1 of conditional formatting for the display previously shown:

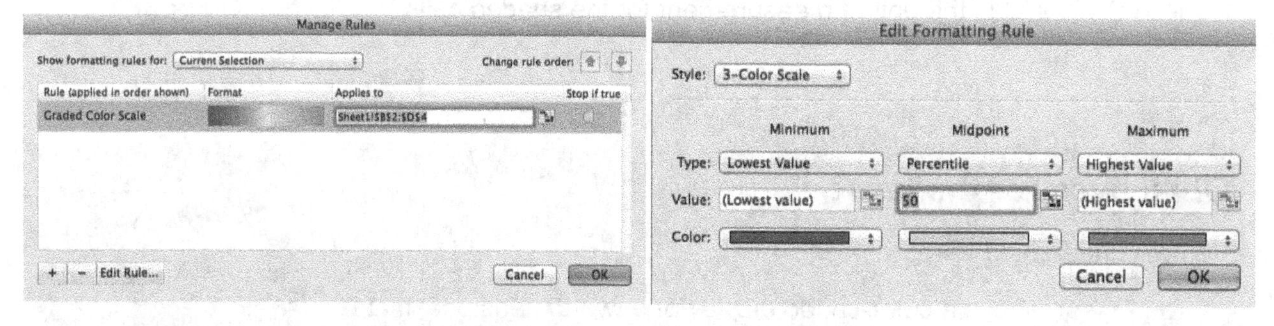

Example 2 of conditional formatting and the resulting display:

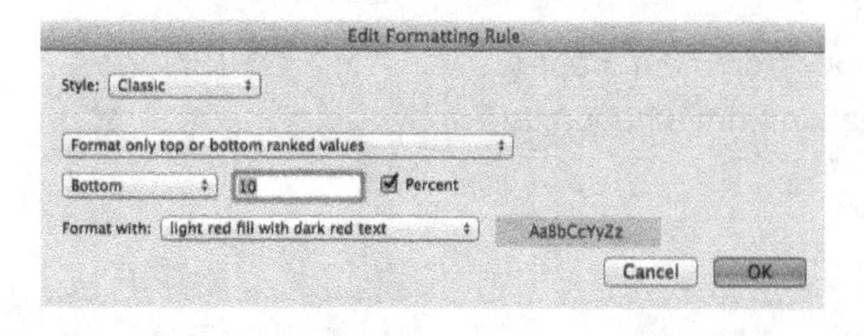

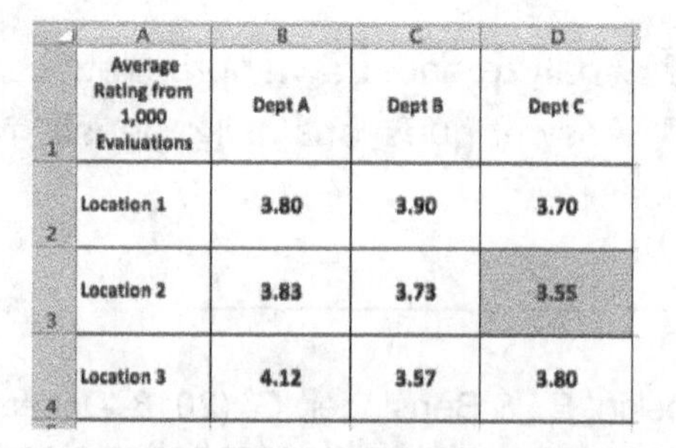

	A	B	C	D
1	Average Rating from 1,000 Evaluations	Dept A	Dept B	Dept C
2	Location 1	3.80	3.90	3.70
3	Location 2	3.83	3.73	3.55
4	Location 3	4.12	3.57	3.80

Part A: What is the range of the ratings for the nine departments?

Part B: Department 3A had a total of 125 ratings. What was the sum of the 125 ratings for this department?

Part C: The nine cells in this heat map are colored using a variable-color scale based on the relative range of values in the nine cells. Some cells will be green (highest values) and some cells will be red (lowest values). Using this variable-color scale, the cells representing the lowest rated departments will be shaded red. Will a cell shaded red always represent a department that needs to improve? How could the color scale be changed so that if the scores are all satisfactory, no red will appear?

3) The figures shown below are heat maps of the student data from a previous lesson. Both figures represent the same student population. The density, in students per square mile, determines the degree of shading of each area. In these two figures, each pixel (each shaded square) is 2,000 feet by 2,000 feet.

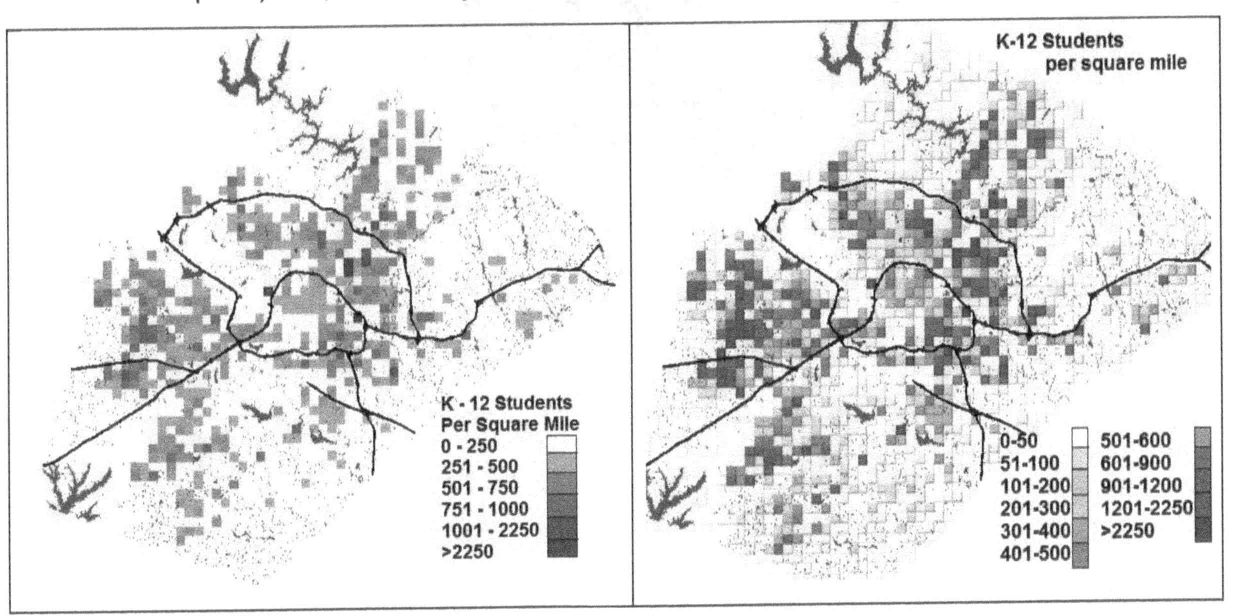

Part A: Compare and contrast the two graphics.

Part B: The pixels in each figure represent a square that is 2,000 feet by 2,000 feet. A mile is 5,280 feet. Choose the closest estimate of the fraction of a square mile that each pixel represents.

a) 50%

b) 25%

c) 12.5%

d) 6.25%

Part C: A circle with a diameter of one mile is drawn in an area of the county with a very dense student population as shown. About how many students are within this circle? Justify your response.

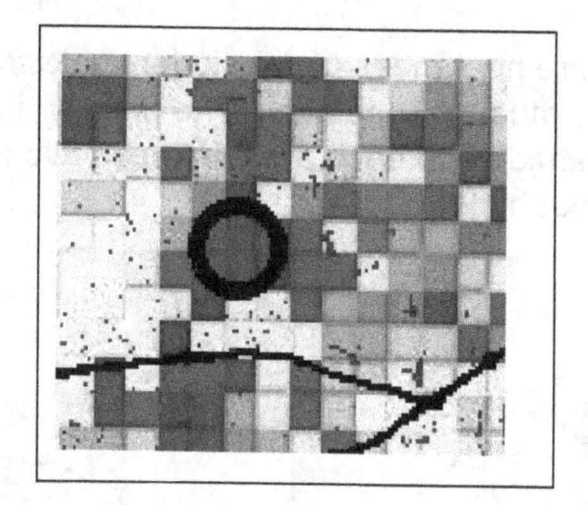

Lesson 17, Part A Decisions, Decisions

In Preview Assignment 17.A, you read about making decisions, such as whether to choose paper or plastic bags at the grocery store. You also considered larger decisions, such as which car to buy.

Remember that each decision may include both quantitative and qualitative components. Recognizing the influence of marketing strategies or personal emotions is an important part of decision making .

Credit: Andres Rodriguez/Fotolia

1) What do you consider when choosing between paper and plastic bags at the grocery store?

Objectives for the lesson

You will understand that:

- ☐ Emotion often plays a role in making financial decisions.
- ☐ Some organizations use psychological approaches to convince consumers to act irrationally.

You will be able to:

- ☐ Use quantitative and qualitative information to make financial decisions.
- ☐ Weigh pros and cons of situations and use that information to make a decision.

In the next set of questions, you will identify numerical (quantitative) and non-numerical (qualitative) information about a cell phone plan, weigh the pros and cons of the options, and come to a reasoned decision.

2) A cell phone company offers the two options listed.

<u>Option 1</u>:

Limited time only!! Save $199.98 and get the new iPhone for only 1¢! (With qualifying 2-year contract and data plan.)

Option 2:

Free yourself from the contract!! Pre-paid nationwide plans for talk, text and web starting at only $35 per month! Wide range of phones available for purchase, including the just-released iPhone at $450.

Part A: What is your first reaction to the plans? Make a snap decision for one of the options. Is one of these options similar to your current phone plan?

Part B: Identify emotional appeals that are in each advertisement.

Part C: Option 1 is an example of a *contract plan*. The price of the phone is very low, but you have to commit to two years of monthly payments for mobile service. In this case, for $75 each month, you get unlimited talk, text and web browsing. Contract plans may also include an activation fee. What are the quantitative and qualitative aspects of this plan that you need to consider in deciding whether to choose the plan? Think about advantages and disadvantages.

Part D: Option 2 is an example of a prepaid phone plan. In this case, you pay the full price for the phone and then you pay a monthly fee based on the plan that you choose. (The $35 quoted in the advertisement is for the minimum plan.) There is usually not an activation fee. Outline the quantitative and qualitative aspects of this plan. What are the advantages and disadvantages of this plan?

Part E: Why would a big phone company market the contract plan more than the prepaid plan?

3) Suppose this month you had an unexpected bill to pay. You may have budgeted your expenses but did not anticipate this additional expense. Suppose the bill is $500 and you need to pay it in the next few days.

Part A: What are your options?

Part B: Let's say that, in this situation, you cannot borrow money from family or friends and you decide to take out a loan. Outline the pros and cons of the three options that follow. Make sure you address quantitative and qualitative aspects of each option.

Option 1: Use your credit card for a cash advance.

To do this, you withdraw cash using your credit card at an ATM. The company will charge a cash-advance fee of 5%, an ATM fee of $3 and 24% annual interest. If you have a balance on your card, the interest automatically is applied to that balance also.

Option 2: Apply for a short-term loan from an instant cash advance company.

This type of loan is often called a "same-day" or "payday" loan. You must write a check or set up an electronic payment for $580 for two weeks from today to pay back the full amount of the loan, plus the interest and fees.

Option 3: Apply for a loan from your bank or credit union.

You fill out the loan paperwork online or in the office. There is a 2% loan origination fee and the annual interest rate is 18%. It may take awhile to get approval.

Part C: Which option do you think you would use? Explain which considerations would be most important to you.

Lesson 17, Part B The Write Approach to Data

Quantitative writing assignments are meant to communicate analyses and interpretations of quantitative data, often to make an argument or support a particular perspective. Numbers are used in a variety of ways to define a problem, to see alternative points of view, to speculate about causes and effects, and to create evidence-based arguments.

In previous math classes, you were probably assigned word problems that were well structured; all the necessary information was provided to find the one correct answer. In contrast, a quantitative writing assignment involves formulating a claim or a conclusion with supporting reasons and evidence.

Credit: BillionPhotos.com/Fotolia

1) In the Preview Assignment, you wrote a paragraph and then assessed yourself with a rubric. What part of the rubric requirements were the most difficult to include in your paragraph?

Objectives for the lesson

You will understand:

- ☐ The components of effective quantitative writing.
- ☐ That writing about data in an organized and concise way can convey information and logical conclusions effectively.

You will be able to:

- ☐ Identify necessary calculations to perform on data and incorporate resulting quantitative information into a summary or analysis.
- ☐ Write brief analyses of data presented in text, table, or graphical form, focusing on key patterns, essential information, and logical conclusions.

We have seen the power of a picture in helping us understand data. In this lesson, we add writing to our list of tools that we use to convey information. Writing is more than just listing facts or writing a couple of sentences. Powerful writing is analytical writing, writing that digs into the data and presents the reader with key patterns and crucial information.

The first step in writing about data is looking at the data and pulling out several key ideas.

Read each paragraph about the graph that you reviewed in Preview Assignment 17.B. Use the Rubric for Quantitative Paragraph Writing to help you critique each paragraph.

2) People are getting older in the United States, especially women. We can see this in the graph because it shows that the ages are going up and up all the time. Maybe this is because of better medicine or money that people have?

3) I see that boys outnumber girls most of the time here until the people get older then it switches. This tells me that women live longer than men. And there are a lot of people too who are 50–54 years old. That's the other interesting thing about this graph. Basically the graph shows that people keep dying and so the population is pretty small when we get to the older people.

4) The graf that I was looking at showed that the population in the US is huge! Looks like there are 10 million girls aged 0–4 and about 10.5 million boys. Then there is a smaller number of girls in the next age group and also for boys. Then it gets a bit biger for both until age 20–24 at around 23 millino total it goes down a bit again. Intersting. At age 40–44 it increases to about 20 million and it keep increasing to the 50–54 year olds, which is about 24 million? Then it goes down for the rest so that by the time we get to 95-99 there's only about 1 million and i can't even figure out the 100+ part since its so small.

5) The United States may be a young country, but its population is not so young. Data from the U.S. Census Bureau show that the largest age group in the U.S. in 2014 is the group of people aged 50–54. This group consists of about 23 million people. The next largest group is the people aged 20–24; this group is almost the same size as the 50–54 year old group. It's also interesting that the U.S. population is fairly evenly distributed among the age groups from 0 to 59. There is a sharp decline at age 60 and the next sets of age groups drop in size by several million at each step. If we think of this historically, the distribution makes sense. The group of people aged 50–54 (and the group just past them, the 55–59 year olds) represents the baby boom, which was a huge population growth in the U.S. after World War II. The other large group is called the baby boomlet and it represents the children of the baby boomers. It will be interesting to see if another bump in population occurs as the children of the baby boomers have their own children.

6) Reflect on quantitative writing and the grading rubric. What is different about this kind of writing?

7) An important step in writing about data is to look at the data and identify several key ideas. You practiced this skill in the Preview Assignment. Let's return to the Census Bureau website, http://www.census.gov/population/international/data/idb/region.php.

   To generate another population pyramid graph, this time select "Afghanistan."

   Part A: Make three quantitative observations about the population distribution for the United States and for Afghanistan.

   Part B: Write a short analysis of the information in these graphs, comparing the population distribution in Afghanistan with the population distribution in the United States. Be sure that your paragraph includes the components outlined in the grading rubric.

Lesson 17, Part C Numbers Never Lie

Just as we are all consumers in the marketplace and need to be careful as we spend money and make choices, we need to be careful when consuming information. We need to look critically at data and graphs. It does not mean that we never trust anyone or any graph, but it does mean that we should be cautious.

Credit: BillionPhotos.com/Fotolia

1) Should you always believe the numbers? Do you think data can lie? Give some examples, either from your life or from earlier in this course or other courses.

Objectives for the lesson

You will understand that:

- ☐ The presentation of data (either graphically or in writing) may be manipulated to tell a particular story.
- ☐ It is essential to critically and accurately evaluate graphs and data, especially when presented in support of an argument.

You will be able to:

- ☐ Find distortions or biases in graphical representations of data.
- ☐ Identify misleading aspects of graphs and mathematical errors in graphs.
- ☐ Write an accurate critical analysis of data, summarize criticisms of data and graphs, and identify potentially misleading information.

2) In this question, we will look at the federal hourly minimum wage from several different perspectives.

   Part A: Has the federal minimum hourly wage always increased or has it ever decreased? What is the current federal minimum wage?

Part B: The graph below shows the historical increases in the federal minimum wage, using data from the Department of Labor.[1] In your group, identify at least two ways that this graph may be distorting the data.

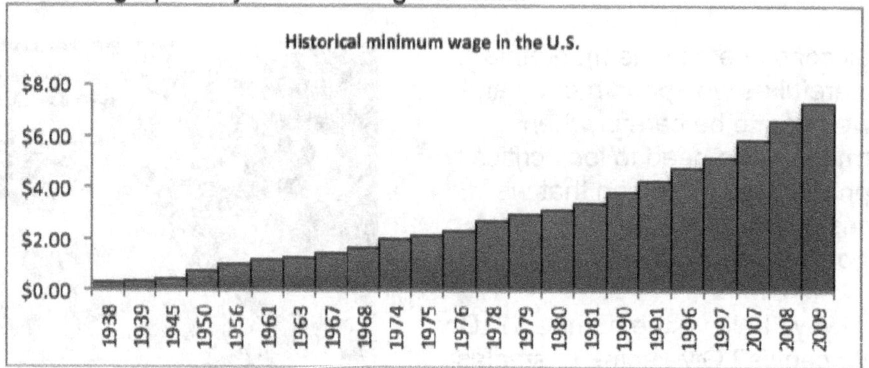

Part C: The graph below shows federal minimum wage for all years between 1938 and 2009. Do you think this graphic display is an appropriate way to show the data? Explain.

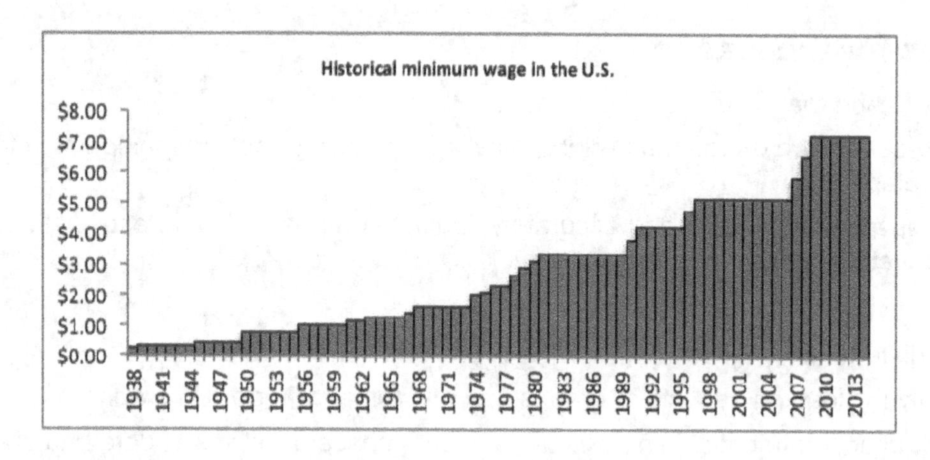

[1] Source: http://www.dol.gov/whd/minwage/chart.htm.

Part D: The graph below shows the minimum wage data with the dollar values adjusted for inflation and converted to 2014 dollars. Why would someone use this graph? What argument does this graph make about the minimum wage?

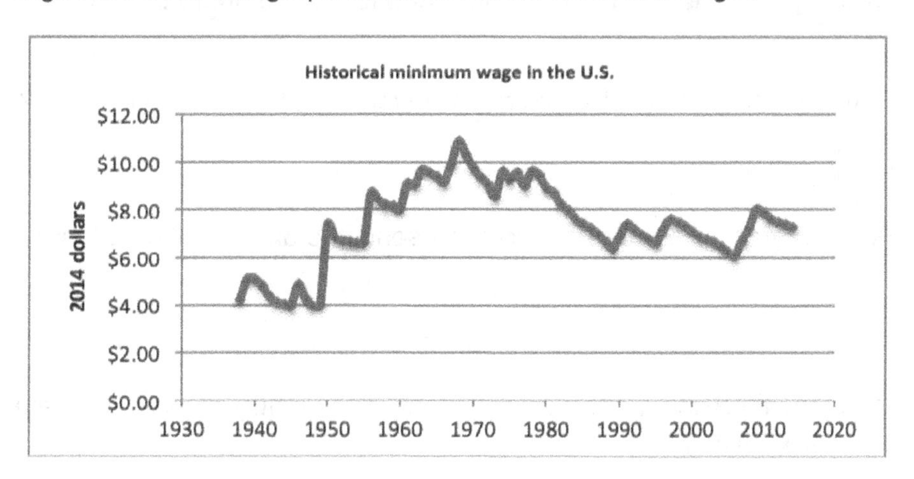

3) A Fox Chicago news story broadcast on Nov. 22, 2009, reported the following about leading Republican candidates for the 2012 Presidential race:

"A new Opinion Dynamics poll for 2012 shows her [Sarah Palin] on top when it comes to landing the nomination. Palin is at 70% . . . Mike Huckabee stands at 63%, Mitt Romney at 60."[2]

The following pie chart was included in the report. What is the problem with this pie chart? How should this data be graphed to be accurate?

[2] Thinkprogressive.org. (2009, November 23.) *Fox Chicago's shows fuzzy math GOP candidate support pie chart.* Retrieved January 4, 2015, from http://youtu.be/-rbyhj8uTT8.

4) Pie charts can be poor representations of data for other reasons. Access “Strange Facts About Cats” at http://www.thepetscentral.com/cats/strange-facts-about-cats. Scroll down until you see the graphic illustrating that cats sleep for 16–18 hours per day.

Part A: Make a quick sketch of the pie charts given. What is wrong with them?

Part B: Sketch a graph that more clearly represents the amount of time that cats sleep.

5) Look at each of the following graphs or statements and identify at least one item that is misleading or a distortion.

Part A: The graph below was created from data from the 2013 Community College Survey of Student Engagement.[3]

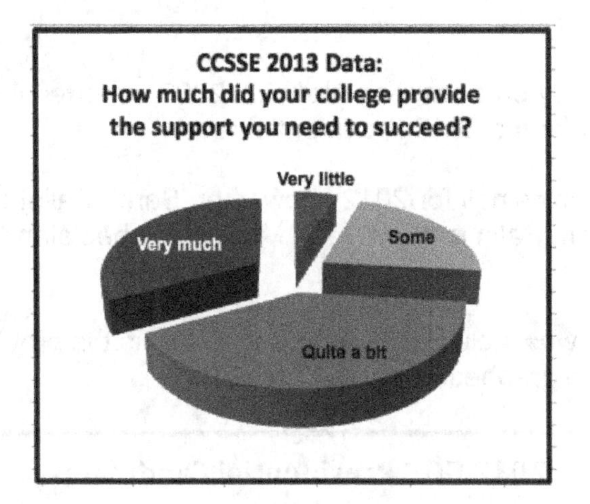

[3] Source:
https://www.ccsse.org/survey/reports/2013/standard_reports/CCSSE_2013_coh_freqs_support_std.pdf

Part B: The text below is from an email message. Do you think this is a real advertisement? What is suspicious about the message?

Learn a Language in 10 Days

(September 2013) Scientific discovery reveals how you can learn a language in just 10 days using this sneaky linguistic secret. This method was recommended by *Forbes* and purchased by the FBI. Shocking results... **FULL STORY**

Part C: Data shown from the U.S. Bureau of Labor Statistics[4] and the Department of Justice Statistics[5] show a strong correlation between fresh fruit consumption and household burglaries. Clearly the old saying "An apple a day keeps the doctor away" should be updated to "An apple a day keeps the burglars away."

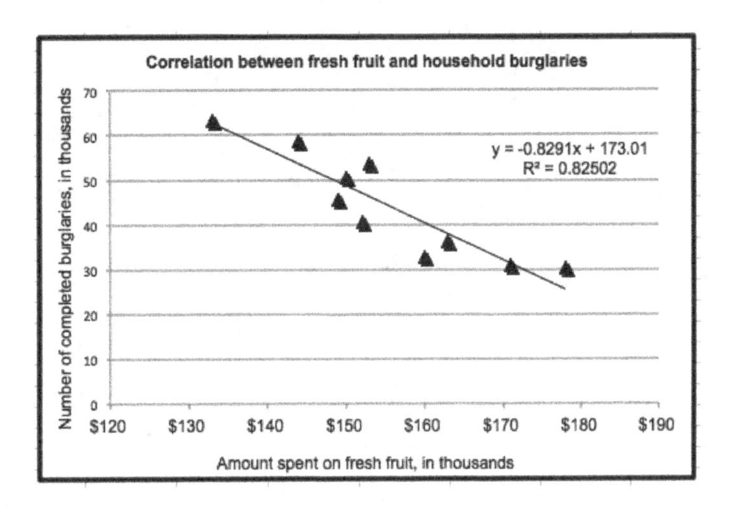

[4] Source: http://www.bls.gov/cex/csxmulti.htm

[5] Source: http://www.bjs.gov/index.cfm?ty=pbdetail&iid=4657

Lesson 17, Part D Can You Feel the Heat?

Let's continue our exploration of long-term trends in climate data. We will see that variations from such trends are to be expected and do not contradict the overall pattern. We will look at data about our own planet Earth.

Scientists are raising concerns about "global warming," saying that the temperature of the Earth is increasing. If the Earth is getting warmer, then why is it still so cold in the winter? Does this mean that global warming is not happening?

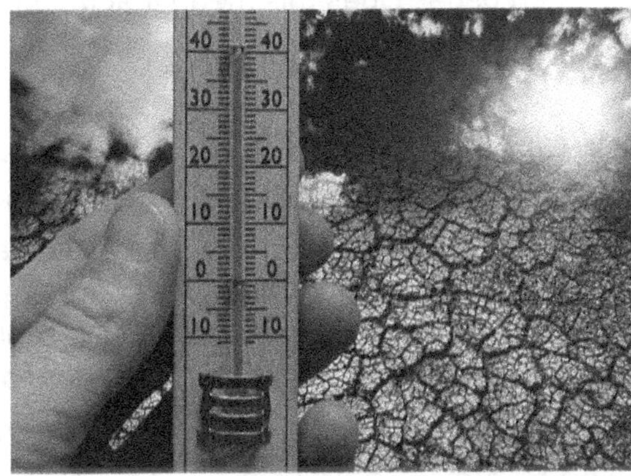

Credit: meryll/Fotolia

1) What do you know about climate change or global warming?

Objectives for the lesson

You will understand that

- ☐ Some issues in our society are complex and involve multiple variables.
- ☐ Analyzing data is important in identifying trends about issues involving several variables.
- ☐ Finding a pattern in data often means analyzing a large set of data, or data values over time, not just an individual value.

You will be able to:

- ☐ Analyze a regression line and use an R and/or R^2 value to determine overall patterns and connections between variables.
- ☐ Make decisions and conclusions based on data, separately from anecdotes or individual experiences.

The White House released its Third National Climate Assessment in May 2014.[1] This report documents the impact of climate change in the United States. The introduction to the report begins with this powerful statement:

---

[1] National Climate Assessment. (2014, May). Climate change impacts in the United States. Retrieved January 4, 2015, from http://nca2014.globalchange.gov.

Climate change, once considered an issue for a distant future, has moved firmly into the present.[2]

In this lesson, you will analyze climate data as a way to understand this complex issue. First, let's look at a concern that has been raised about climate change.

2) The winter in 2013–2014 was the coldest since 2002. The summer of 2014 was also colder than usual. Does this mean that global warming is not happening?

3) Open Spreadsheet_17D, which contains data from www.climate.gov. The first tab is a spreadsheet that contains information about the average global temperature between 1880 through 2013. Note that these temperatures are in degrees Celsius, not Fahrenheit. Scientists typically use the Celsius scale. In this scale, 0 degrees is the freezing point of water, and 100 degrees in the boiling point of water. Each row of the spreadsheet contains a year and the difference between the average temperature of that year and the long-term average world temperature.

Part A: Make a column graph of the data.

Part B: Discuss any trends that you see. Do these data support the hypothesis that the Earth is experiencing global warming?

Part C: Go back to your graph and change the chart type to "Scatter." Then use the spreadsheet menus to fit a trendline to the graph. Make sure you include the equation of the trendline and the R^2 value.

Part D: What does the trendline equation tell you about the average change in the Earth's temperature over time? Be specific.

4) Click on the tab "TempCarbonDioxide" on your spreadsheet. This worksheet shows data about the levels of carbon dioxide in the Earth's atmosphere and temperature data from the previous worksheet. The carbon dioxide abundance, as it is called, is measured in parts per

---

[2] Source: http://nca2014.globalchange.gov./highlights/overview/overview#intro-section-2

million, and the measurements were made at the National Oceanic and Atmospheric Administration's observatory in Mauna Loa, Hawaii.

Part A: Use the spreadsheet to make a scatterplot of the temperature and carbon dioxide data, including a trendline.

Part B: What trend do you see? How strong is the correlation between these variables?

Part C: Discuss why the correlation between the two main variables is so strong. As you know, correlation is not the same as causation. What could be causing the climate changes? Do you think that humans are largely responsible for recent changes in the world's climate? Think back to the information you learned in the Preview Assignment about the greenhouse effect.

Lesson 18, Part A Tornado Climatology Mini-Project

In Preview Assignment 18.A, you analyzed a graph and read an article about U.S. tornado climatology. Now let's extend our exploration of tornado data.

1) The table shows the ten deadliest tornadoes in U.S. history through 2014.[1]

Date	State(s)	Injuries	Deaths
3/18/1925	MO, IL, IN	2,027	695
5/6/1840	LA, MS	109	317
5/26/1896	MO, IL	1,000	255
4/5/1936	MS	700	216
4/6/1936	GA	1,600	203
4/9/1947	TX, KS, OK	970	181
5/22/2011	MO	1,000	158
5/24/1908	LA, MS	770	143
6/12/1899	WI	200	117
6/8/1953	WI	844	116

Part A: How many of the tornadoes in the table occurred within the past decade?

Part B: Does the data suggest that tornadoes were more prevalent and more powerful in the past than now? Explain.

Part C: What other ways (besides injuries and deaths) are people affected by tornadoes?

Objectives for the lesson

You will understand that:

- ☐ Graphics and statistics may mislead readers/viewers.

You will be able to:

- ☐ Use proportions to reason and make interpretations.
- ☐ Choose appropriate ways to represent data and information in an effort to represent the complete and accurate story.
- ☐ Write an informative, objective report that is appropriate for the audience.

---

[1] Source: http://www.ncdc.noaa.gov/climate-information/extreme-events/us-tornado-climatology/deadliest

The map you analyzed in the Preview Assignment has been colored to differentiate the states by the average number of tornadoes each year in that state.[2]

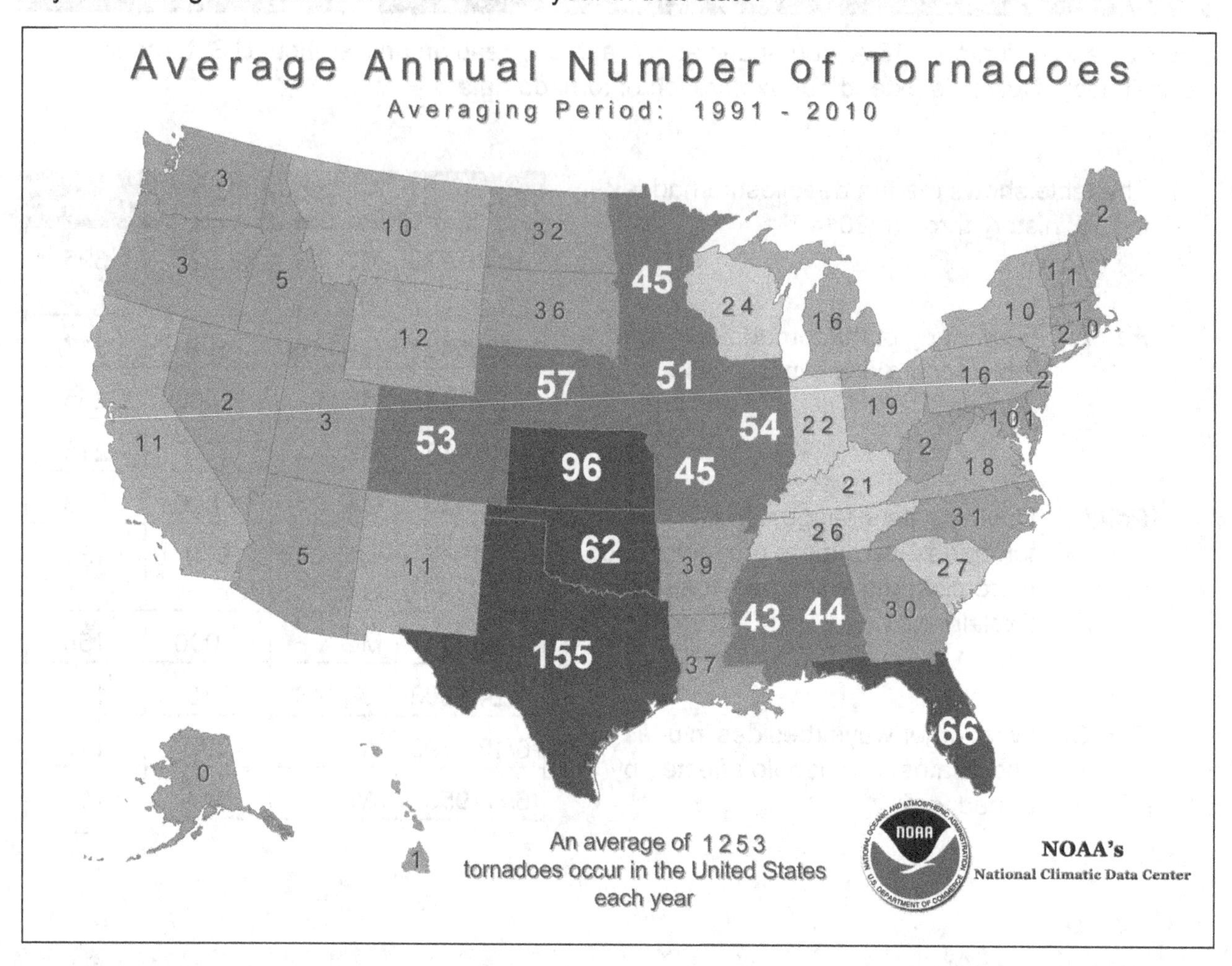

[2] Source: http://www.ncdc.noaa.gov/climate-information/extreme-events/us-tornado-climatology

2) Suggest an appropriate range of average annual number of tornadoes that may have been used for each color in the following coloring scheme.

Color	Average Annual Number of Tornadoes
Red	
Orange	
Pink	
Beige	
Bright Green	
Gray Green	

3) Suppose someone asserts that the graphic is misleading. Give a rationale to support that claim.

4) What other information would be helpful if your group wanted to portray the likelihood of someone in a given state being affected by a tornado in any given year?

Lesson 18, Part B The Making of a Model

A **mathematical model** can be described as a mathematical structure that approximates the features of a phenomenon. That is, a mathematical model describes a system using mathematical concepts and language (an equation, a graph, a description, a spreadsheet, etc.).

In previous lessons on modeling, you were usually told which model would be appropriate for the given situation. This lesson is the first in a series in which you will learn strategies for choosing an appropriate mathematical model for data.

1) Work your way through the process shown in the graphic. Describe what happens as you move through the process.

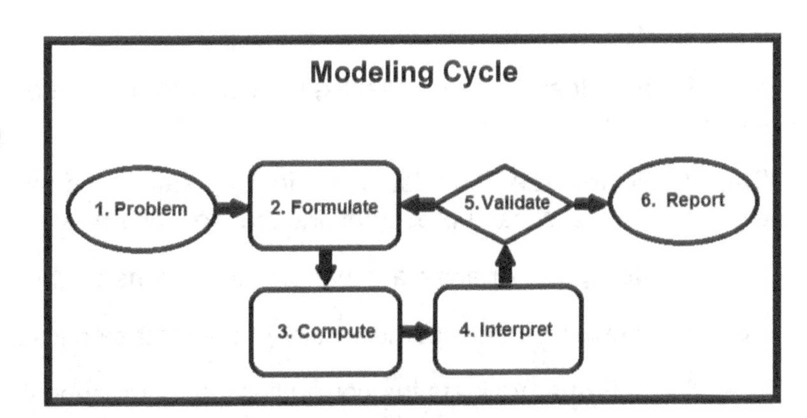

Objectives for the lesson

You will understand:

- ☐ The steps of a modeling cycle and the limitations on models.

You will be able to:

- ☐ Develop, test, and justify a model for a physical phenomenon.

A basic modeling cycle with six steps is shown above.[1] In developing and creating a mathematical model, it is important to address each step. Of special note is the "validate" step, as it is critical to validate and check that your model and equations are correct and that the model is accurate.

Also keep in mind, as usual, the importance of being able to communicate and justify your results

[1] The Modeling Cycle in this lesson is adapted from a process outlined in AMATYC's *The Right Stuff: Appropriate Mathematics for All Students* project, http://www.therightstuff.matyc.org/module0/Module0.html.

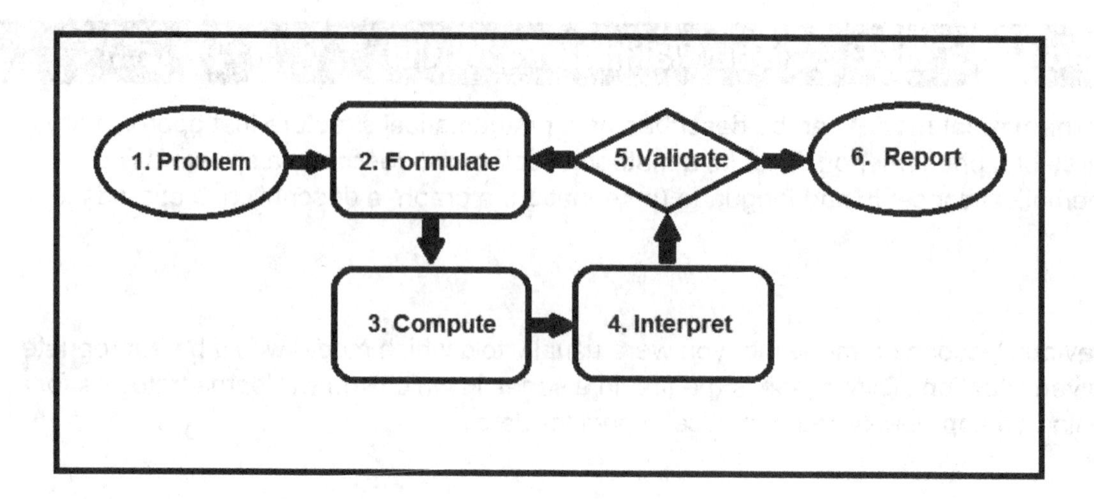

Step 1 – Problem: Identify variables in the situation and select those that represent the essential features.

Step 2 – Formulate: Formulate a model by creating and selecting geometric, graphical, tabular, algebraic, or statistical representations that describe relationships between the variables.

Step 3 – Compute: Analyze and perform operations on these relationships to draw conclusions.

Step 4 – Interpret: Interpret the results of the mathematics in terms of the original situation.

Step 5 – Validate: Validate the conclusions by comparing them with the situation, and then either improve the model or, if it is acceptable, report the conclusions.

Step 6 – Report: Report the conclusions and the reasoning behind them. Include choices, assumptions, and approximations that were required.

2) Stadium seating can be constructed using concrete blocks. The figure below shows how the rectangular blocks can be arranged to provide rows of seating.

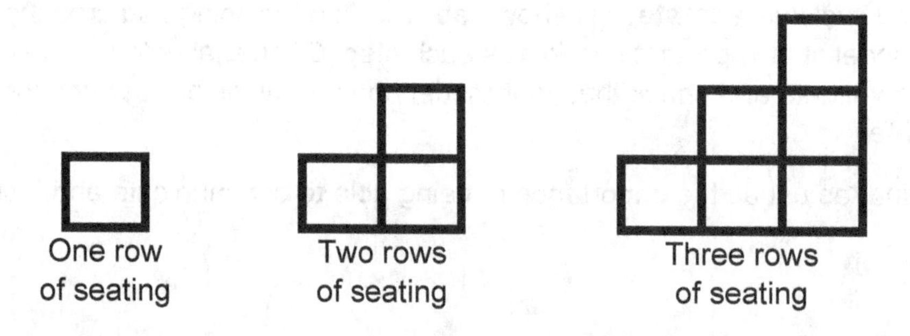

Work your way through each of the six steps of the Modeling Cycle to develop a model that determines the number of blocks required to have eight rows of seating. Be prepared to share and explain your model.

3) How confident are you that the model you created in question 2 can be used to find the exact number of blocks required to build r rows of stadium seating? Explain your answer.

Lesson 18, Part C What a Wonderful World!

The total surface area of the Earth is about 200 million square miles. About 71% of the surface area is water. There are about 7.1 billion people living on Earth.[1]

Credit: wayne_0216/Fotolia

1) What were your observations after viewing the video about the growth in world population since 1 AD?

Objectives for the lesson

You will understand that:

- ☐ Data from real-world models can create linear, exponential, logistic, and periodic models.

You will be able to:

- ☐ Employ the steps of the Modeling Cycle and determine the model that best fits the data.

The table and graph below show the approximate world population at various time points since 1 AD.

	A	B
1	Year	World Population (approximate)
2	1	170,000,000
3	500	190,000,000
4	1000	310,000,000
5	1500	425,000,000
6	1700	790,000,000
7	1804	1,000,000,000
8	1850	1,260,000,000
9	1900	1,650,000,000
10	1927	2,000,000,000
11	1960	3,000,000,000
12	1974	4,000,000,000
13	1987	5,000,000,000
14	1999	6,000,000,000
15	2012	7,000,000,000

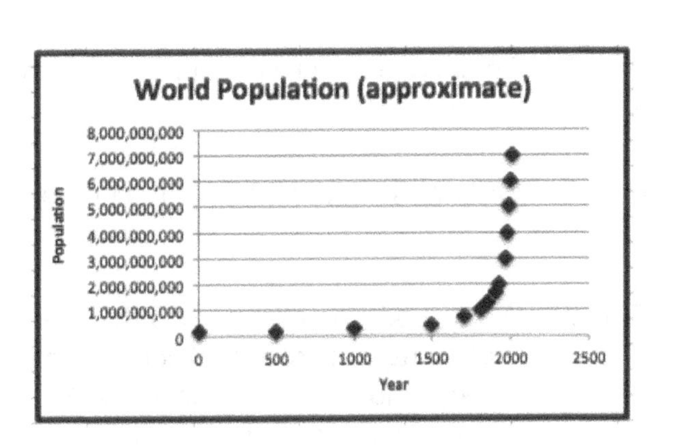

---

[1] World Bank. (2013). *World Development Indicators*. Retrieved January 6, 2015, from www.worldbank.org.

2) Which model would best fit the data: linear, exponential, logistic, periodic, or other? Justify your choice (think about the steps in the Modeling Cycle).

3) The table and graph below give a different view of the history of the approximate world population (in billions). In this case, the estimates are at various time points since 1650.

	A	B
1	Year	World Population (approximate, in billions)
2	1650	0.47
3	1750	0.69
4	1850	1.1
5	1900	1.6
6	1930	2.1
7	1940	2.3
8	1950	2.4
9	1960	3
10	1970	3.7
11	1980	4.4
12	1985	4.8
13	1990	5.28
14	1995	5.7
15	2000	6.1
16	2005	6.5
17	2010	6.89
18	2014	7.1

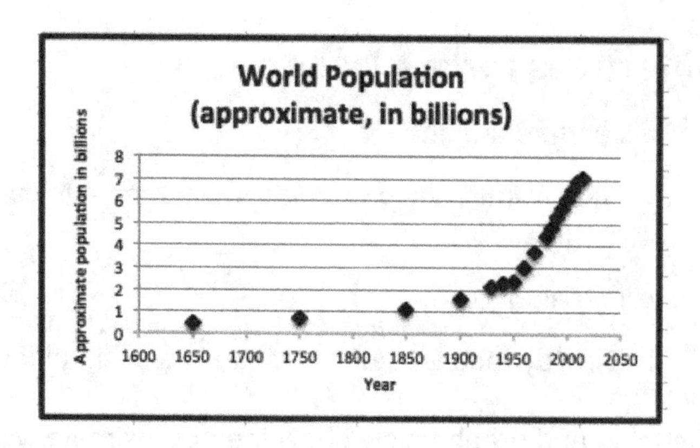

Part A: On average, what was the annual growth in population between 1930 and 1940?

Part B: On average, what was the annual growth in population between 1970 and 1980?

Part C: What was the average annual percentage growth during these two time periods?

Part D: When modeling phenomena that are not governed by laws of nature or physics, it is often necessary to add conditions to the model, that is, restrictions. In the case of the world population since 1 AD, an exponential model can approximate the population rather well. However, it would not be reasonable to use that model for predicting the world population in the next several decades, or even centuries, as conditions prior to 1950 do not have as much to do with future trends as do more recent events, more recent trends, and more recent demographics.

There is an obvious pattern change around 1950—it is more obvious from the graph than the table. What happened during that time that you think may have contributed to such a drastic change?

4) Let's concentrate now on the population growth following that period of change in the 1950s.

Part A: Complete the following table by calculating the average annual change and average annual percentage change.

Year	World Population (approximate, in billions)	Average Annual Change (millions of persons per year)	Average Annual Percentage Change
1960	3		
1970	3.7		
1980	4.4		
1985	4.8		
1990	5.28		
1995	5.7		
2000	6.1		
2005	6.5		
2010	6.89		
2015[2]	7.3		

Part B: Which model would best fit the population of the world since 1960: linear, exponential, logistic, periodic, or other? Justify your choice (think about the steps in the Modeling Cycle).

---

[2] Source: http://populationpyramid.net/world/2015

Lesson 18, Part D Mathematical Models

In Preview Assignment 18.D, you read and answered questions about the Eiffel Tower in Paris, France. Our exploration of population growth and modeling takes us back to Paris and now to two other world cities: Shenzhen, China, and Lagos, Nigeria.

Credit: ryanking999/Fotolia

1) If you could choose a 10-day vacation in one of these countries, which one would you choose? Why?

Objectives for the lesson

You will understand that:

- ☐ A mathematical model of a real-world scenario used to make a forecast may be complex and comprised of multiple models.

You will be able to:

- ☐ Choose and create an appropriate algebraic model and make a reasonable forecast of the population of a city.

The next three questions present specific population information about the cities of:

2) Paris, France

3) Shenzhen, China

4) Lagos, Nigeria

Your instructor will describe the process for deciding which question you will complete.

Remember the steps of the Modeling Cycle as you work on the problem and prepare a population forecast. Be prepared to report your results to the class.

Modeling Cycle

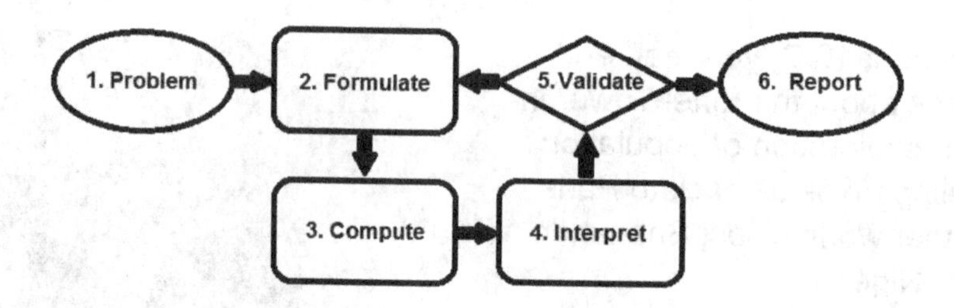

2) **Paris, France**

France is divided into administrative regions (similar to states).[1] The most populous is the Paris Region, also called the Île-de-France, which covers 2,845 square kilometers. (The city of Paris covers a much smaller area within the region.) The population of this region at various times is given in the following table.

Year	Paris Region Population (in millions)
1975	8.55
1982	8.71
1990	9.32
1999	9.64
2010	10.46

Part A: In order to predict the population of this area in 2020, would a linear model or an exponential model be more reasonable? Be prepared to justify your conclusion.

Part B: Create the model you chose in Part A and use it to predict the population of the Paris Region in 2020. How confident are you in your prediction?

[1 1] Source: Institut national de la statistique et des études économiques, www.insee.fr

3) **Shenzhen, China**

There are a number of replicas of the Eiffel Tower around the world, including one at the Window of the World theme park in Shenzhen, China.

Located north of Hong Kong, the city of Shenzhen was originally a small village in the province of Guangdong in southern China. Rapid growth took place in the late 20th century and Shenzhen is now a major city.[2]

The population of Shenzhen at various times since 1979 is given in the following table.

Year	Shenzhen Population (in millions)[3]
1979	0.314
1982	0.351
1990	1.215
1995	4.491
2000	7.008
2005	8.277
2010	10.358
2011	10.467

Part A: In order to predict the population of this area in 2020, would a linear model or an exponential model be more reasonable? Be prepared to justify your conclusion.

Part B: Create the model you chose in Part A and use it to predict the population of Shenzhen in 2020. How confident are you in your prediction?

[2] Source: Collegiate Journal of Anthropology, http://anthrojournal.com/issue/october-2011/article/but-at-what-cost-shenzhen-china-and-the-social-implications-of-urban-success1

[3] Source: http://www.sztj.gov.cn/nj2012/szcn/3-1_1ch.htm

4) **Lagos, Nigeria**

Lagos, Nigeria has also seen incredible growth over the past 20 years. Although it remains largely impoverished with many slums, it has seen steady improvement in its governance; enhanced public transportation, and a better overall quality of life. The emergence of this urban area may point the way to a strategy for cities to save countries.[4]

The population of Lagos at various times since 1950 is given in the following table.

Year	Lagos Population (in millions)[5]
1950	0.33
1960	0.76
1970	1.4
1980	2.6
1990	4.8
2000	7.3
2010	11

Part A: In order to predict the population of this area in 2020, would a linear model or an exponential model be more reasonable? Be prepared to justify your conclusion.

Part B: Create the model you chose in Part A and use it to predict the population of the Lagos in 2020. How confident are you in your prediction?

[4] Kaplan, Seth D. (2014, January 7). What makes Lagos a model city? *The New York Times*. Retrieved January 10, 2015, from http://www.nytimes.com/2014/01/08/opinion/what-makes-lagos-a-model-city.html?_r=0.

[5] Source: http://data.worldbank.org/indicator/EN.URB.LCTY

Resource **Overview**

Resource pages and their relationship to assignments

Each assignment in the course will be structured around the principles of:

- Developing skills and understanding.
- Making connections to prior learning.
- Preparing for future lessons.
- Reading, writing, and reflection.

Each assignment will contain problems in which you practice the skills from the lesson, including extending those skills in a new way or applying them to a new situation.

Taking control of your own learning

In addition, each assignment will include questions designed to prepare you for future lessons. You will be given resources to help you, if needed.

When you use videos as a tool for refreshing on concepts, it is best to actively engage with the material in the video, rather than watching passively. For example:

- Copy the examples.
- Pause the video and work problems when directed to do so.
- When you have to correct your work, do so underneath the original work and write an explanation of your error and/or the correction.
- Watch the video a second time and add comments by the examples.

Self-regulating your learning

One goal of this course is to increase your ability to learn efficiently and effectively. This means learning faster and learning smarter—what scientists call being a "self-regulated learner." The following section explains what this means.

Self-regulating your learning means you ***plan*** your work, monitor your ***work*** and progress, and then ***reflect*** on your planning and strategies and what you could do to be more effective. These are the three phases of Self-Regulated Learning (SRL). They are introduced below and will be followed up on later in the course.

Plan: Before doing a problem or assignment, self-regulated learners ***plan***. They think about what they already know or do not know, decide what strategies to use to finish the problem, and plan how much time it will take. Research has shown that math experts often spend much more time planning how they will do a problem than they do actually completing it. Novices, the people who are just starting out, often do the opposite.

Work: Self-regulated learners use effective strategies as they ***work*** to solve problems. They actively *monitor* what study strategies are working and make changes when they are not working. When they do not know which strategy would be better, they ask for help. Self-regulated learners also keep themselves focused while they are working and pay attention to their feelings to avoid getting frustrated.

Reflect: Usually after an assignment or problem is done, self-regulated learners take time to reflect about what worked well and what did not. Based on that reflection, they think about how to change their approach in their future. The ***reflect*** phase helps self-regulated learners understand more about how they learn so they can become more efficient and more effective the next time. Reflecting is important for doing a better job next time you plan for a new problem or assignment.

You can think of these three phases as a cycle. You incorporate what you learned during the reflect stage in your *next* plan phase, making you a more effective learner as you repeat this process many times. The most effective students get in the habit of working this way:

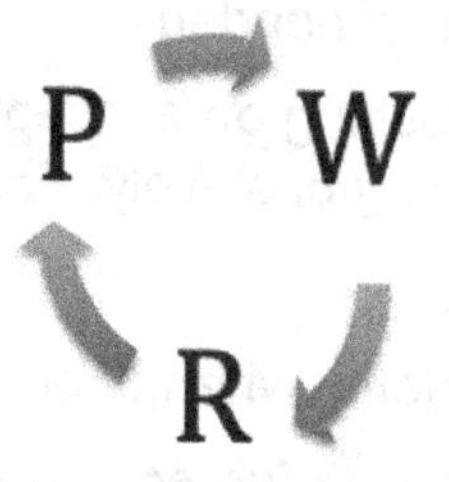

For most people, self-regulating takes time, practice, and hard work, but it is always possible. People can improve even if, in the beginning, they did not self-regulate their learning very well. The more you practice something and the more you train your brain to think in certain ways, the easier it becomes.

Resource 5-Number Summary and Boxplots

5-number summary

In your previous course, you learned about the **5-number summary** for describing a set of data. Consider the following set:

550, 61, 75, 228, 79, 121, 79, 129, 240, 150, 147, 72, 142, 50

Step 1: Put the values in order from smallest to largest and identify the smallest value (the minimum) and the largest value (the maximum).

50, 61, 72, 75, 79, 79, 121, 129, 142, 147, 150, 228, 240, 550

The minimum value is 50 and the maximum value is 550.

Step 2: Find the median.

50, 61, 72, 75, 79, 79, 121,   129, 142, 147, 150, 228, 240, 550

↑

The median is (121 + 129)/2 = 125

Step 3: Find Q1 and Q3.

Find the median of the smaller half of the numbers:

50, 61, 72, 75, 79, 79, 121

Q1 is 75 ↑

Find the median of the larger half of the numbers:

129, 142, 147, 150, 228, 240, 550

Q3 is 150 ↑

Step 4: Report your 5-number summary in order.

Minimum	50
Q1 (first quartile)	75
Median	125
Q3 (third quartile)	150
Maximum	550

Box-and-whisker plots: Graphical displays of the 5-number summary

Box-and-whisker plots (often called simply "boxplots") are a graphical way of illustrating data in quartiles (groups of 25% of the data).

A basic box-and-whisker plot can be created using the steps described below.

Step 1: Create a number line with evenly spaced increments, which reaches below the minimum value and above the maximum value of the data set.

50 100 150 200 250 300 350 400 450 500 550

Step 2: Place vertical markers above each of the values in the 5-number summary from the previous page.

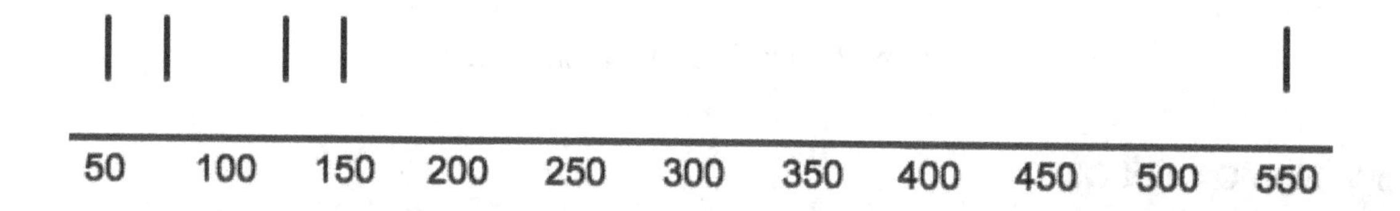

Step 3: Draw horizontal markers to create the box and the whiskers as shown.

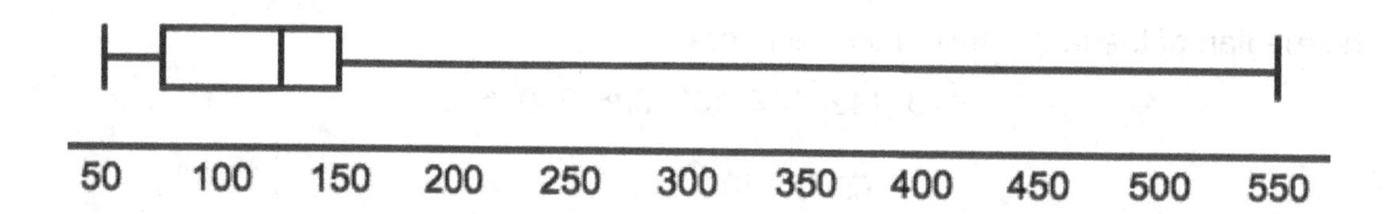

Step 4: Analyze the result.

Sample observation: Notice that the whisker on the right side is very long; this means that the top 25% of the data values are very spread out. The left whisker is very short; the lowest 25% of the data values are very close together.

Remember, this is the basic boxplot format. More sophisticated boxplots can be used to highlight unusual values in the data.

Resource **Algebraic Terminology**

Some notes about algebraic terminology

Formulas are a type of an **algebraic equation**. You have probably seen algebraic equations in previous math classes. For example:

$$y = x + 3$$

Each side of this equation is called an **algebra expression**.

$$\overbrace{y}^{expression} = \overbrace{x+3}^{expression}$$

So "$x+3$" is an expression and "y" is an expression. The equal sign indicates that the two expressions are equal, thus forming an **equation**.
Therefore, an equation must have an equal sign with expressions on each side.

Expression = Expression

[Note that this is exactly the same as $x+3=y$]

An equation defines a **sequence of calculations**, often using algebra to shorten the information. In the example above, this sequence is:

1) Start with x
2) Add three to x
3) The result is y

Notice how much shorter $y=x+3$ is than the three listed steps.

The word **formula** is usually used to express important and non-changing relationships, especially in contexts such as science, business, medicine, sports, or statistics. For example, in Lesson 11, you used the formula for the area of a rectangle, $A = L \bullet W$. This is a formula because the relationship between area and the length and width of a rectangle is always the same.

An example of an equation would be if you had a job in which you make $12 per hour. This relationship could be written algebraically as $P=12h$ where P is your pay in dollars, and h is the number of hours you work. If you get a raise, the relationship would change. You also might call the equation a **model** because it models a situation using mathematics.

Resource **Coordinate Plane**

Vocabulary

Axes: A coordinate plane has two axes that measure distance in two dimensions. The *horizontal axis* goes from left to right. In previous classes, you may have called this the *x-axis*. The *vertical axis* goes up and down. This is sometimes called the *y-axis*. The axes are two number lines that create a grid on the coordinate plane. **Note:** *Axis* is singular and *axes* is plural.

Origin: The point at which the two axes intersect or cross is called the origin. This point represents 0 for both axes. To the left of this point, the horizontal axis is negative; to the right, it is positive. Below the origin, the vertical axis is negative; above the origin, it is positive. You can see this in the numbers along each axis. These numbers are called the *scale*.

Ordered pair: Each location or point on the coordinate plane is defined by an ordered pair. You can think of this as the "address" of a point. Ordered pairs are written in a set of parentheses. They are called *ordered pairs* because they must contain two numbers and the order of the numbers is important. The first number is the distance and direction going left or right from the origin, and the second number is the distance and direction going up or down. The ordered pair for the origin is (0, 0).

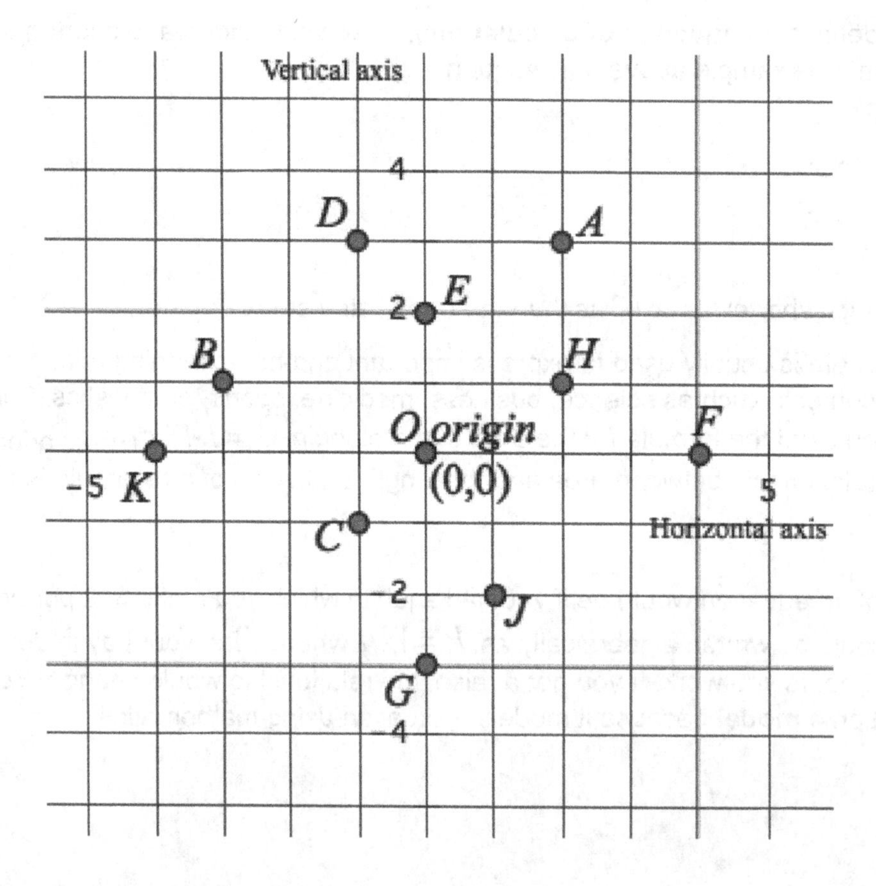

Follow these steps to find the point represented by the ordered pair (2, 3):

<u>Step 1</u> First, think about the address of the point. If this were a street address, the ordered pair tells you to walk 2 blocks horizontally in the positive direction (right) and then walk 3 units vertically in the positive direction (up).

<u>Step 2</u> Start at the origin. Go 2 units to the right because this is the positive side of the horizontal axis.

<u>Step 3</u> Go 3 units up.

Point A on the graph is the point (2, 3). A few other examples from the graph are given below:

Point B: (–3, 1) Point E: (0, 2) Point F: (4, 0)

Where would you place point P (–4, 5)?

Resource Dimensional Analysis

Quantitative reasoning skill: Ratios and unit rates

Unit rates are ratios with a denominator of 1, although they are not always written as fractions.

For example, 60 mph is the same as $\frac{60 \text{ miles}}{1 \text{ hour}}$.

The language "miles per hour" implies that the operation is miles divided by 1 hour.

As another example, in 2012, the federal minimum wage was $7.25/hour. This means that an employee earns $7.25 for 1 hour of work, or $\frac{\$7.25}{1 \text{ hour}}$.

A worker may also be paid a weekly salary. If an advertisement states that a job pays $320 for a 40-hour work week, then that position can be compared to the previous job by converting to a unit rate:

$$\frac{\$320}{40 \text{ hours}} = \frac{8 \cdot 40}{1 \cdot 40} = \frac{\$8}{1 \text{ hour}}$$

The second job pays better.

Another way to think about the calculation above is as a division problem: $320 \div 40 = 8$

This can be helpful when the numerator and denominator do not have a common factor.

Quantitative reasoning skill: Conversion factors

In the above example, the fraction was simplified by dividing out the common factor of 40/40 (which is equivalent to 1). A fraction that is a ratio of quantities can be equivalent to 1 even when the numerator and denominator are not the same number. However, it is necessary that the numerator and denominator represent equivalent quantities. For example, the following fractions are all forms of one:

$$\frac{16 \text{ ounces}}{1 \text{ pound}} \qquad \frac{1 \text{ mile}}{5{,}280 \text{ feet}} \qquad \frac{60 \text{ minutes}}{1 \text{ hour}}$$

These types of ratios are sometimes called conversion factors because they can be used to covert between units.

The example below shows how to set up a multiplication problem with the rate and the conversion factor to convert miles per hour to miles per minute.

$$\frac{35 \text{ miles}}{1 \boxed{\text{hour}}} \cdot \frac{1 \boxed{\text{hour}}}{60 \text{ minutes}}$$

Notice that the conversion factor is written so that the units of hours are in the numerator. This is because you want the *hours* label to divide out in the same way that common factors divided out in the weekly salary problem above. This leaves the units of miles/minute as shown here:

$$\rightarrow \frac{35 \text{ miles}}{1 \cancel{\text{hour}}} \cdot \frac{1 \cancel{\text{hour}}}{60 \text{ minutes}} \rightarrow \frac{35 \text{ miles}}{60 \text{ minutes}} \rightarrow \frac{0.58 \text{ mile}}{1 \text{ minute}}$$

Before continuing, make sure you can answer the following question:

- How was the 0.58 calculated?

Quantitative reasoning skill: Dimensional analysis

Dimensional analysis, unit analysis, or unit conversion are all names for the process of using conversion factors to set up and solve certain types of problems. Many professionals—including pharmacists, dietitians, lab technicians, and nurses—use unit analysis. It is also useful for everyday conversions in cooking, finances, and currency exchanges. Many people can do simple conversions without dimensional analysis; however, they will likely make mistakes on more complex problems.

The advantage of using dimensional analysis is that it is a way to check your calculations. While it is always important that you develop your own methods to solve problems, this is a time when you are encouraged to learn and use a specific method. Once you have learned dimensional analysis, you can decide when to use it and when to use other methods.

In Lesson 1, Part C, you needed to convert inches into feet and then into miles. The computation below shows how dimensional analysis can help you organize your work.

Example 1:

$$\frac{1{,}000 \text{ people}}{1{,}500 \text{ feet}} \times \frac{5{,}280 \text{ feet}}{1 \text{ mile}} =$$

Unit labels can "cancel" in the same way that common factors do.

$$\frac{1{,}000 \text{ people}}{1{,}500 \cancel{\text{feet}}} \times \frac{5{,}280 \cancel{\text{feet}}}{1 \text{ mile}} =$$

$$\frac{5{,}280{,}000 \text{ people}}{1{,}500 \text{ mile}} = 3{,}520 \text{ people per mile}$$

Multiply numerators.

Multiply denominators.

Simplify.

Remember that you must set up your conversion factor (the multiplier) so that matching labels appear in one numerator and one denominator.

Some more extensive examples are shown on the next page.

Example 2:

Here is an example converting 1 year into seconds:

$$\frac{365 \text{ days}}{1 \text{ year}} \times \frac{24 \text{ hours}}{1 \text{ day}} \times \frac{60 \text{ minutes}}{1 \text{ hour}} \times \frac{60 \text{ seconds}}{1 \text{ minute}} =$$

$$\frac{365}{1 \text{ year}} \times 24 \times 60 \times 60 \text{ seconds} = 31{,}536{,}000 \text{ seconds per year}$$

Example 3:

A nurse or a pharmacist might need to know how many tablespoons are in a 250-mL (milliliter) bottle of medication. In this case, we are converting metric units (milliliters) to U.S. units (tablespoons).

There are 250 mL in one bottle, and a quick Internet search indicates 1 tablespoon is about 14.79 milliliters.

$$\frac{250 \text{ mL}}{1 \text{ bottle}} \times \frac{1 \text{ tablespoon}}{14.79 \text{ mL}} =$$

$$\frac{250 \text{ mL}}{1 \text{ bottle}} \times \frac{1 \text{ tablespoon}}{14.79 \text{ mL}} =$$

$$\frac{250}{14.79} \approx 16.9 \text{ tablespoons per bottle}$$

Example 4:

A father found some instructions on the Internet for building a treehouse. The instructions were in metric units but he only had a standard English ruler. The instructions said the boards for the framing should be 2.5 meters long. How many inches should the dad measure?

$$\frac{2.5 \text{ meters}}{1 \text{ board}} \times \frac{100 \text{ cm}}{1 \text{ meter}} \times \frac{1 \text{ inch}}{2.54 \text{ cm}} =$$

$$\frac{2.5 \text{ meters}}{1 \text{ board}} \times \frac{100 \text{ cm}}{1 \text{ meter}} \times \frac{1 \text{ inch}}{2.54 \text{ cm}} =$$

$$\frac{250}{2.54} \approx 98.4 \text{ inches per board}$$

Of course, if you can connect to the internet, you could just search for “free online conversion calculator”!

Resource **Equivalent Fractions**

Quantitative reasoning skill: Equivalent fractions

Two fractions are equivalent if they have the same value or represent the same part of an object.

For example, the figure shows that 1/2, 2/4, and 4/8 all represent the same part of a whole. They are equivalent fractions.

Recall that the *denominator* of a fraction represents the number of parts into which the whole has been divided. The *numerator* represents a count of the number of parts.

So, $\frac{4}{8}$ means that the whole is divided into 8 equal parts, and 4 of these parts are counted.

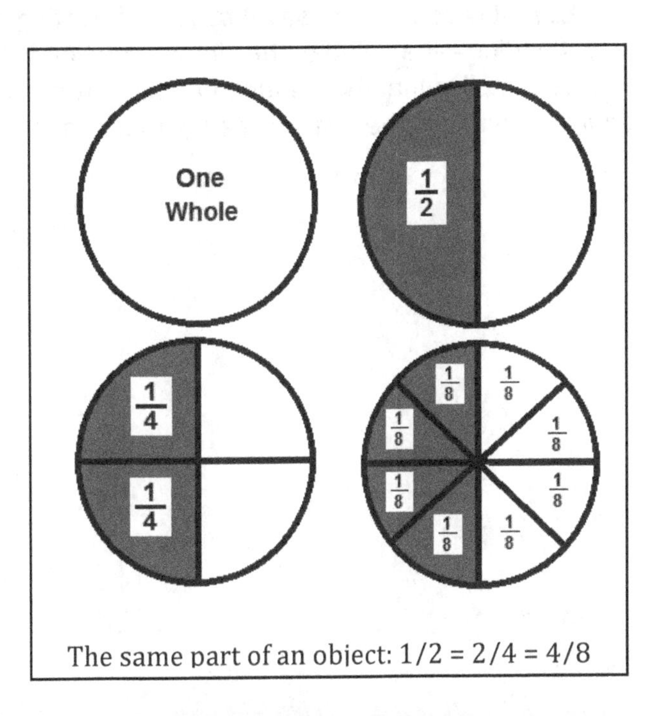

The same part of an object: 1/2 = 2/4 = 4/8

Quantitative reasoning skill: Simplifying fractions

The fraction 50/100 is equivalent to 1/2. Note that you can write:

$$\frac{50}{100} = \frac{1 \cdot 50}{2 \cdot 50} = \frac{1}{2} \cdot 1$$

The above calculation shows that both 50 and 100 can be written as a number times 50. You say that 50 and 100 have a "common factor of 50."

Another way to think of this is that the number 1 (written as $\frac{50}{50}$) is embedded in the fraction $\frac{50}{100}$.

"1" is a special number in mathematics because if you multiply any number by 1, you get a result that is equivalent to the original.

By dividing $\frac{50}{50}$ to get 1, you simplify $\frac{50}{100}$ to $\frac{1}{2}$.

In this case, the word *simplify* means that the fraction has been written in an equivalent form with smaller numbers. The *simplest* form means that the fraction is written using the smallest possible numbers. In general, answers should always be given in simplest form unless the question specifically calls for a different form.

Caution: It is common to say that you are writing the fraction in "reduced form." This language is misleading—the value of the simpler fraction is the same as the original fraction, but the word *reduced* implies that the "reduced fraction" represents a smaller quantity. The terminology *simplest form* or *lowest terms* makes more sense.

Resource Four Representations of Relationships

Mathematical relationships can be represented in four ways: models (equations), tables, graphs, and verbal descriptions. If you are struggling with a problem, approaching it with a different representation may help you to make sense of the work.

Model or equation

In Lesson 12, your class wrote a mathematical equation for the relationship of the price of gas and the cost of driving Jenna's car. An equation is useful because it can be used to calculate the cost values. As you saw with the formula for braking distance in Lesson 13, equations are also useful for communicating complex relationships. In writing equations, it is always important to define what the variables represent, including units. For example, in Lesson 12, the variables were defined as shown below. Note that each definition includes what the variable represents, such as *cost of Jenna's car*, and the units in which this quantity is measured, such as *$/mile*.

J = Cost of Jenna's car in $/mile

g = Price of gas ($/gal)

These variables were used in the mathematical equation, $J = \frac{g}{22} + 0.146$.

Table

Another way that you could have represented this relationship between the price of gas and the cost of driving the car is in a table that shows values of g and J as *ordered pairs*. An ordered pair is two values that are matched together in a given relationship. You used this representation in Lesson 13 when you explored how one variable affected another. Tables are helpful for recognizing patterns and general relationships or for giving information about specific values. A table should always have labels for each column. The labels should include units when appropriate.

Price of Gas ($/gal)	Cost of Driving Jenna's Car ($/mile)
3.00	0.28
3.50	0.31
4.00	0.33
4.50	0.35

Graph

In the last few assignments of Lesson 10, Part A through Lesson 14, Part D, you practiced with graphs. A graph provides a visual representation of the situation. It helps you to see how the variables are related to each other and make predictions about future values or values in between those in your table. The horizontal and vertical axis of the graph should be labeled, including units.

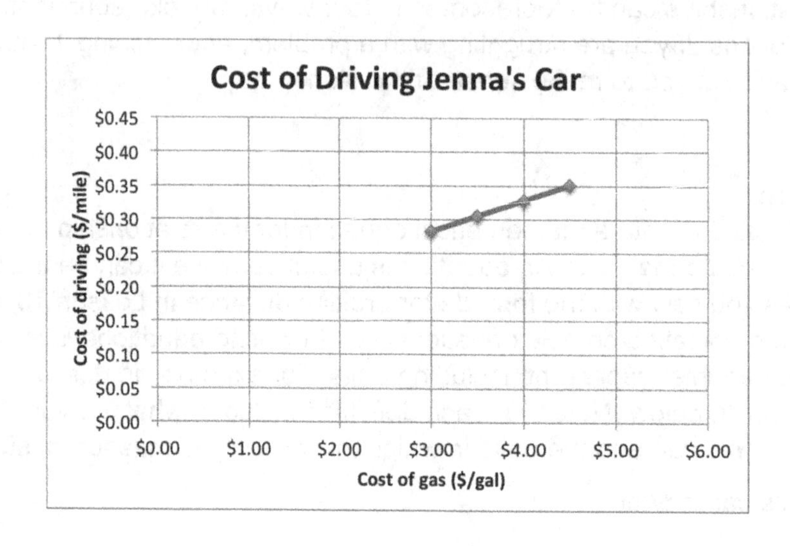

Verbal description

A verbal description explains the relationship in words, which can help you make sense of what the relationship means in the context.

For $J = \frac{g}{22} + 0.146$, the fraction $\frac{g}{22}$ represents the per-mile cost of the gas, which depends on the price of the gas $\frac{\$}{\text{1 gallon}} \bullet \frac{\text{1 gallon}}{\text{22 miles}} = \frac{\$}{\text{mile}}$ and 0.146 represents the per-mile costs associated with oil changes, tire wear, etc. So the equation $J = \frac{g}{22} + 0.146$ represents the total per-mile cost of Jenna's car. Verbally, Jenna might say, "I need to find the per-mile cost of my car so that I can compare it to the cost of a car rental. Dividing the gas price by 22 miles per gallon will give me the per-mile cost of the gas and then I need to add in the scaled costs of maintenance to get the total."

Summary

Throughout this course, you have learned that having the skill to move between different forms and tools is important in problem solving. Alternating among the four representations of mathematical relationships is another example of this. In some cases, you may struggle writing an equation, but find that starting with a table is helpful. You might want a graph for a visual representation, but also need to express a relationship in words. It is important that you can translate one form into another and also that you can choose which form is most useful in a specific situation.

Resource **Fractions, Decimals, Percentages**

Language of fractions, decimals, and percentages

There are several important vocabulary words you should know and use.

- A ratio is a comparison of two numbers by division. You will see many different types of ratios in this course. Some ratios are a special type called a percentage. A percentage is a ratio because it is a number compared to 100.
- Percentages are a relationship between two values: the *comparison value* and the *base value*. The relationship is described as a *percentage rate,* which is shown with a percentage symbol (%). This indicates that the rate is out of 100.

  Example: 10 is 20% of 50.

  10 is the comparison value.

  50 is the base value.

  20% is the percentage rate; it can be written as a decimal by using the relationship to 100: $\frac{20}{100} = 0.2$

- Fractions have two parts: $\frac{\text{numerator}}{\text{denominator}}$
- Every fraction can be written in *equivalent* forms (e.g., $\frac{2}{3} = \frac{4}{6} = \frac{6}{9}$). It is often useful to write the fraction in the form with the smallest numbers. This is called *simplified* or *reduced*. In the example, $\frac{2}{3}$ is in simplest form.

On the next page are some **common percentage benchmarks**.

Simplified Fraction	Percent	Decimal
$\frac{1}{100}$	1%	0.01
$\frac{1}{10}$		
		0.2
	25%	
$\frac{1}{3}$	round to the nearest one percent	round to nearest hundredth
		0.5
$\frac{2}{3}$	round to the nearest one percent	round to nearest hundredth
		0.75

Calculating percentage rates

Additional examples of finding the percent of a number and calculating percentage rate can be found at http://www.purplemath.com.

Resource Length, Area, and Volume

Length

Length is one-dimensional. An example would be the length of an extension cord that you need to plug in an electronic device. Examples of units of measure for length are inches, feet, yards, or miles (or in the metric system, centimeters, meters, or kilometers).

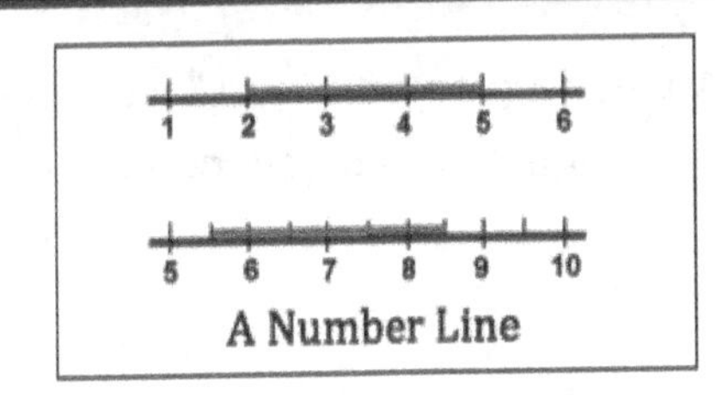

A Number Line

A number line can be used to model lengths.

The thicker segment on each number line shown above is 3 units long. If the scale is in inches, each line segment is 3 inches long. If the scale is in feet, each line segment is 3 feet long.

Area

Area is two-dimensional and is measured in square units. The total number of one-foot square tiles needed to cover the floor of a room is an example of area measured in square feet, and can be modeled with a rectangle. Recall the formula for the area of a rectangle:

$$A = L \times W$$

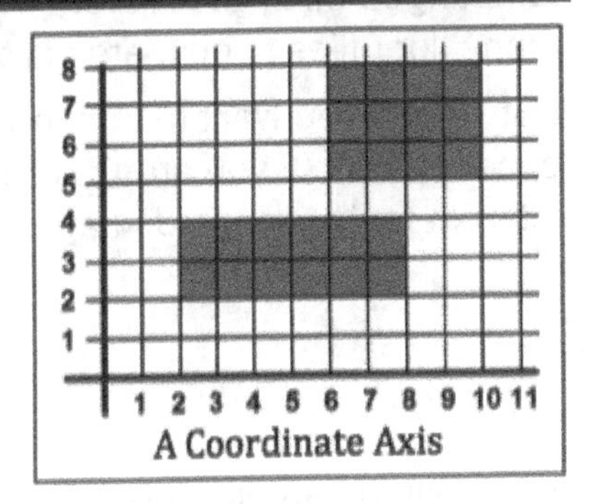

A Coordinate Axis

The area of a rectangle is the product of the length and the width, which is a shortcut for counting the number of square units needed to cover the rectangle.

Each of the two shaded areas on the coordinate axis has an area of 12 square units. If the horizontal and vertical scales are in inches, each area is 12 square inches. If the scales are in feet, each area is 12 square feet. Notice that the regions measured do not have to be squares, yet the area is measured in square units.

If the units are in inches, the area of the top rectangle is:

$$\begin{aligned} A &= (3\ inches) \times (4\ inches) \\ &= (3 \times 4) \times (inches \times inches) \\ &= 12\ inches \times inches \\ &= 12\ square\ inches \end{aligned}$$

If the units are in feet, the area of the bottom rectangle is:

$$\begin{aligned} A &= (2\ feet) \times (6\ feet) \\ &= (2 \times 6) \times (feet \times feet) \\ &= 12\ feet \times feet \\ &= 12\ square\ feet \end{aligned}$$

For more about labeling units, see the note on the next page.

Note 1:

It is common to abbreviate the units of measure using exponents. If the area is $A = 12\ feet \times feet = 12\ square\ feet$, we often write $A = 12\ ft^2$

Notice the connection to algebra here!

Multiplying $(3 feet)$ by $(4 feet)$ is similar to multiplying $(3x) \times (4x)$. You multiply the numbers in front of the variables (coefficients), and then multiply the variables:

$$(3x) \times (4x) =$$
$$(3 \bullet 4)(x \bullet x) =$$
$$12x^2$$

Note 2:

It is common to confuse length and area formulas. Look at the bottom rectangle. Because it is shaded, it is tempting to think about area. If you want to know how many floor tiles to buy, area is the correct concept.

But what if you want to trim the edges of the room with baseboards? (If you aren't sure about this term, search the internet for "baseboard images.")

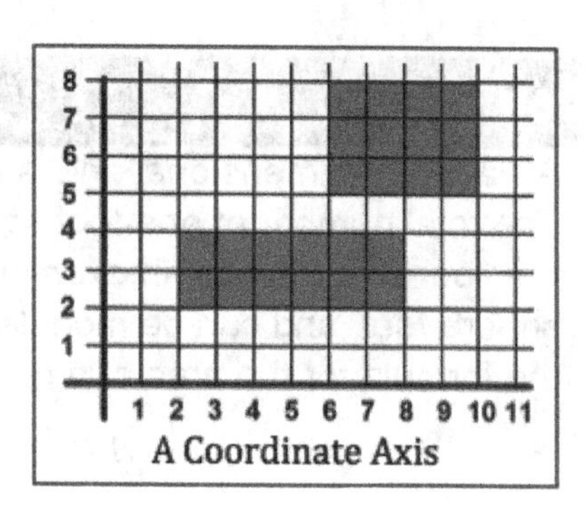

A Coordinate Axis

The length of the distance around a shape is called the **perimeter**. To calculate this length, simply add the total number of units as if traveling around the edge. For example, if the units are in feet, then the length of the line around the bottom rectangle is

$$P = 2 feet + 6 feet + 2 feet + 6 feet$$
$$= 16\ feet$$

The arithmetic operation for length is **addition**, and the unit of measure is feet (you are adding up a lot of feet, so the final result is feet). By comparison, the arithmetic operation to compute area is **multiplication**, and the unit of measure is square feet (you are determining the number of square tiles). Again, this connects to algebra. To add algebraic terms, you must have **like terms**, meaning terms with the same variables:

$$2x + 6x + 2x + 6x =$$
$$16x$$

You cannot add $2x + 3y$ just as you cannot add $2\ feet + 3\ inches$.

Not every shape you need to find the length or the area of will be a square or rectangle. Think about a circular rug in the living room, or a gazebo in the shape of an octagon.

Circle length and area

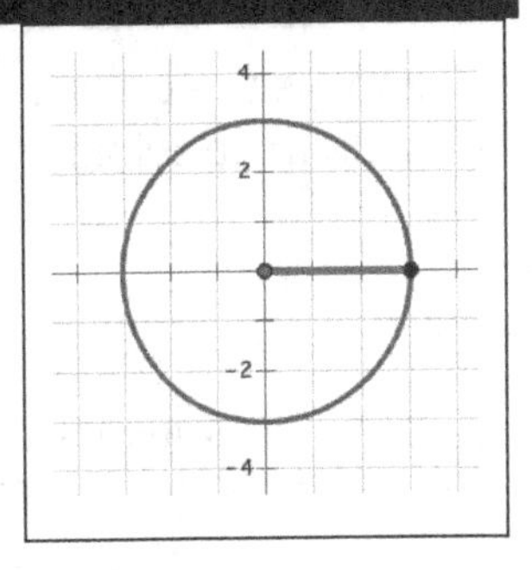

The distance around a circle is called the **circumference**, which can be found with the formula:

$$C = 2\pi r$$

The area of a circle is given by the formula:

$$A = \pi r^2$$

- π is a *constant* that is approximately 3.14159
  (You probably learned 3.14, but carrying additional decimal places reduces the amount of rounding error.)
- r is the radius of the circle, which *varies* depending on the size of the circle. It is the distance from the center to any point on the circle.
- C is the circumference of the circle, which *varies* depending on the radius.
- A is the area of the circle, which *varies* depending on the radius.

In this example, the radius is 3 units. Let's say those units represent inches.

The circumference is:

$$\begin{aligned} C &= 2\pi r \\ C &= 2\pi(3\ inches) \\ &= 2 \cdot 3 \cdot \pi\ inches \\ &= 6\pi\ inches\ \ (exactly) \\ &\approx 6 \cdot 3.14159\ inches \\ &\approx 18.85\ inches \end{aligned}$$

Look at the length of 1 unit on the radius. Does $19\ inches$ seem like a reasonable estimate for the distance around the circle?

The area is:

$$\begin{aligned} A &= \pi r^2 \\ A &= \pi(3\ inches)^2 \\ &= \pi(3\ inches) \times (3\ inches) \\ &= \pi \cdot 3 \cdot 3(inches)(inches) \\ &= 9\pi\ square\ inches \\ &\approx 9 \cdot 3.14159\ square\ inches \\ &\approx 28.27\ in^2 \end{aligned}$$

Look at the grid. Is $28\ in^2$ a reasonable estimate of the area of the circle?

Volume

Volume is three-dimensional and is measured in cubic units. The formula to calculate the volume of a box is:

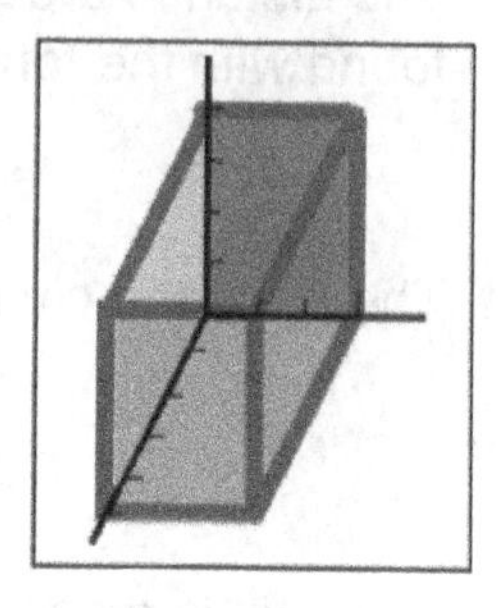

$$V = L \times W \times H$$

If the graph at the right is in inches, then this shape is 5 inches long, 3 inches wide, and 4 inches high.

$$V = (5\ inches) \times (3\ inches) \times (4\ inches)$$

$$V = 60\ inches^3\ or\ 60\ cubic\ inches$$

Resource **Mean, Mode, and Median**

The **sum** of a set of numbers is the result obtained by adding the values in the set.

The **size** of a set of numbers is *the number of numbers* in the set and is often designated as "***n***."

Example:

The set *A* represents the amounts Joey spent when she used her debit card yesterday. How many transactions did Joey have, and what is the total amount she spent?

$A = \{38, 14, 12, 26\}$

The size of the set is $n = 4$ Joey had 4 transactions.

How did you find ***n***? Simply count the number of data values!

The sum of the set is $38 + 14 + 12 + 26 = 90$. Joey spent $90.

How did you find this? You may have used your calculator. However, you could challenge yourself to work mentally. Sometimes regrouping the numbers makes them easier to work with. Look again at the set *A*. Do you see a way to rearrange the set to make pairs of numbers you could add together more easily in your head? (Check the bottom of this page if you are stuck.)

You can regroup the numbers in your head as follows:

$38 + 14 + 12 + 26 =$

$(38 + 12) + (14 + 26) =$

$50 + 40 =$

90

Averages

People often talk about "averages," and you probably have an idea of what is meant by that. Now you will look at more formal mathematical ways of defining averages. In mathematics, you call an average a **measure of center** because an average is a way of measuring or *quantifying* the center of a set of data. There are different measures of center because there are different ways to define the center.

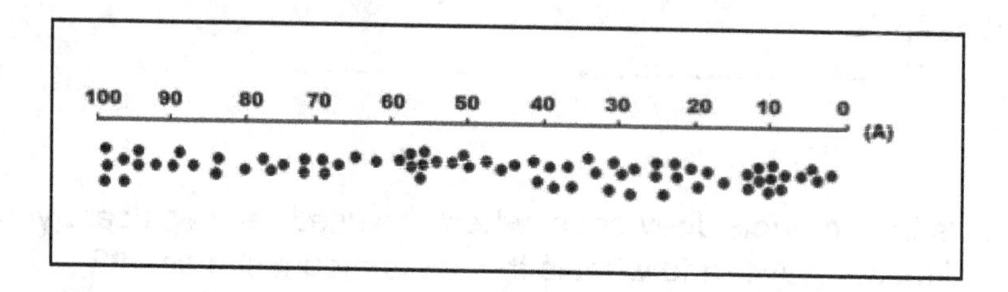

Think about a long line of people waiting to buy tickets for a concert. (Figure A shows a line about 100 feet long, and each dot represents a person in the line.) In some sections of the line, people are grouped together very closely, while in other sections of the line, people are spread out. How would you describe where the center of the line is?

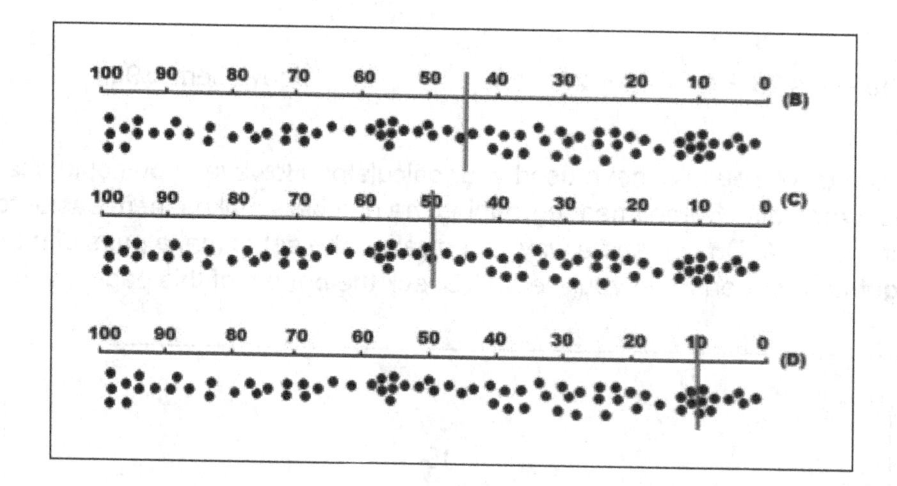

- Would you define the center of the line by finding the point at which half the people in the line are on one side and half are on the other (see Figure B)?
- Is the center based on the length of the line even though there would be more people on one side of the center than on the other (see Figure C)?
- Would you place the center among the largest groups of people (see Figure D)?

The answer would depend on what you needed the center for. When working with data, you need different measures for different purposes.

Different types of averages are described on the next page.

Mean (Arithmetic Average)

Find the average of numeric values by finding the sum of the values and dividing the sum by the number of values. The mean is what most people mean when they say "average."

Example:

$$\begin{aligned}\text{Mean} &= \frac{\text{sum}}{n}\\ &= \frac{X1+X2+X3+X4+X5}{n}\\ &= \frac{32+12+8+42+100}{5}\\ &= \frac{194}{5}\\ &= 38.8\end{aligned}$$

Manny's Purchases

	W	X
1	Shirt	32
2	Hat	12
3	Lunch	8
4	Gas	42
5	Gift Card	100

So, there are three ways to talk about this value:

1) The set is "centered" at 38.8.
2) The mean of the numbers is 38.8.
3) Manny spent an average of $38.80 per purchase.

Mode

Find the mode by finding the number(s) that occur(s) most frequently. There may be more than one mode.

Example 1:

Find the mode of 18, 23, 45, 18, 36.

The number 18 occurs twice, more than any other number, so the mode is 18.

Example 2:

Find the mode of the quiz grades. 70, 75, 75, 75, 80, 80, 85, 85, 85, 90, 95, 95

The number 75 occurs three times, as does 85. This is more than any other number, so there are two modes, 75 and 85.

Median

Find the median of a set of numbers by first arranging the data in order of size.

1) If there is an odd number of values, the median is the middle number.
2) If there is an even number of values, the median is the mean of the two middle numbers.

Example *(data set with odd number of values)*
To find the median of Manny's purchases, write the numbers in order: 8, 12, 32, 42, 100

There are 5 values (an odd number), so the median is the number in the middle.

$$8,\ 12,\ \boxed{32},\ 42,\ 100$$

The median is 32.

Example *(data set with even number of values)*
To find the median of Joey's purchases, write the numbers in order: 12, 14, 26, 38

With an even number of values, there is no one middle number. Find the median by finding **the mean of the two middle numbers**:

$$\text{Median} = \frac{14+26}{2} = \frac{40}{2} = 20$$

Resource **Multiplying and Dividing Fractions**

Quantitative reasoning skill: Multiplying fractions

Another way to think about fractions is in terms of area. Look at the rectangle below. The fraction $\frac{2}{3}$ can be represented by the dark gray area found by dividing the rectangle into thirds horizontally and shading 2 of the sections as shown.

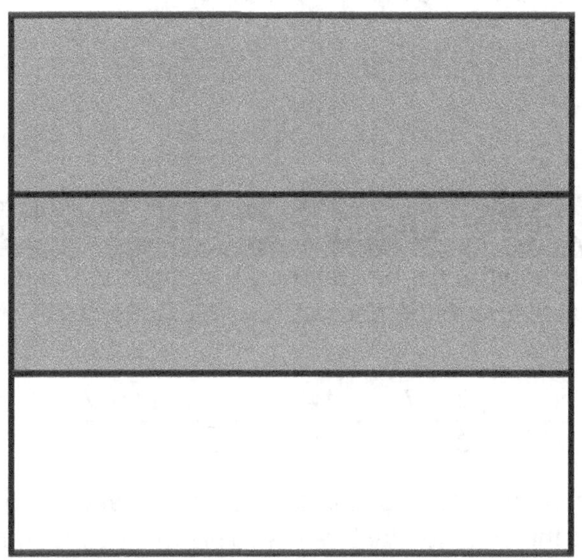

You can also think about **multiplying fractions** in terms of area of a rectangle.

Shade $\frac{2}{3}$ as indicated above. Now represent $\frac{4}{5}$ by dividing the rectangle vertically into fifths and shading 4 of the 5 sections.

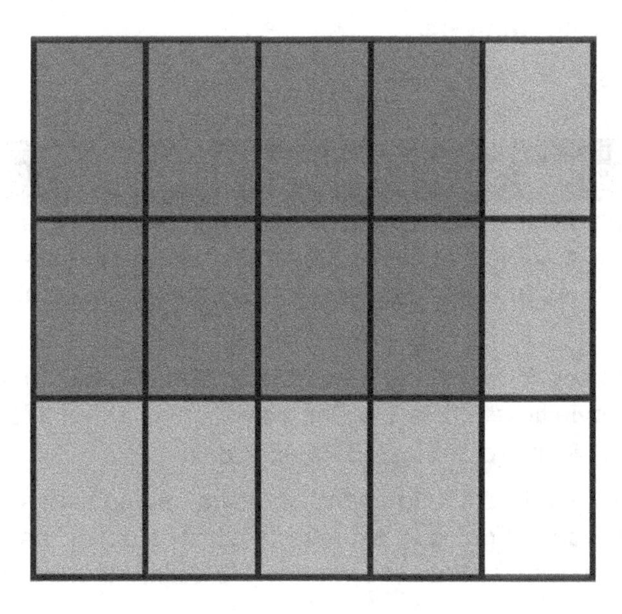

- The product $\frac{2}{3} \cdot \frac{4}{5}$ can be represented by the region that was shaded twice.
- Notice that the rectangle is now divided into 15 regions ($15 = 3 \times 5$), and the number of those regions that are dark gray is 8 ($8 = 2 \times 4$).
- So 8 out of 15 pieces are dark gray, or 8/15.
- This prompts a rule for multiplying fractions: multiply the numerators ($2 \cdot 4$) and multiply the denominators ($3 \cdot 5$), and simplify if possible.
- Therefore, $\frac{2}{3} \cdot \frac{4}{5} = \frac{8}{15}$.

Quantitative reasoning skill: Simplifying before multiplying fractions

The fact that common factors in the denominator and numerator of a number can be divided to make 1 can be used to make multiplying fractions easier. Consider the following multiplication problem.

$$\frac{2}{3} \cdot \frac{7}{8} \rightarrow \frac{14}{24} \rightarrow \frac{7 \cdot 2}{12 \cdot 2} \rightarrow \frac{7}{12}$$

If you see that there is a common factor of 2 in the numerator and denominator before multiplying, you can divide the common factors first. This makes the multiplication easier because you have smaller numbers to work with, and the simplification is complete.

$$\frac{2}{3} \cdot \frac{7}{8} \rightarrow \frac{2}{3} \cdot \frac{7}{4 \cdot 2} \rightarrow \frac{\not{2}}{3} \cdot \frac{7}{4 \cdot \not{2}} \rightarrow \frac{1 \cdot 7}{3 \cdot 4} \rightarrow \frac{7}{12}$$

This is an important concept when working with ratios with units. You will learn more about this in the Resource **Dimensional Analysis**.

Quantitative reasoning skill: Dividing fractions

Many people struggle with dividing fractions because it is difficult to visualize. A full explanation of the mathematics behind dividing fractions is beyond what the authors can do in these materials. Instead, the authors are providing you with a context that might help you remember how to divide fractions.

Suppose you have $48 to spend on going to the movies during a month. How many tickets can you buy in a month? A movie ticket costs $8. One way to think about this is that you want to know how many groups of $8 there are in $48, or $48 \div 8$.

In the same way, suppose you had $10 to spend on downloading songs for a 1/2 dollar. (For the sake of the mathematics, you are going to express "a half of a dollar" as a fraction instead of as a decimal.) This means you want to know how many 1/2 dollars there are in $10. Your common sense probably tells you that the answer is 20 because every 1 dollar has two halves. So you multiplied 10×2. Look at this written as a calculation:

$$10 \div \frac{1}{2} \textit{ is the same as } 10 \bullet \frac{2}{1}$$

$$\frac{1}{2} \textit{ and } \frac{2}{1} \textit{ are called reciprocals.}$$

So you say that division is the same as multiplying by the reciprocal. Here are more examples:

$$4 \div \frac{1}{2} \rightarrow 4 \bullet \frac{2}{1} \rightarrow 8$$

$$12 \div \frac{2}{3} \rightarrow 12 \bullet \frac{3}{2} \rightarrow \frac{36}{2} \rightarrow 18$$

$$\frac{4}{5} \div 2 \rightarrow \frac{4}{5} \bullet \frac{5}{2} \rightarrow \frac{20}{10} \rightarrow 2$$

Resource **Number-Word Combinations**

Combining our large number work with our rounding work can make large numbers much easier to work with in certain problems where exact values are not needed. We do this by approximating large numbers with a **number-word combination**. Consider the following examples:

25,145,561 can be rounded to the nearest million as 25,000,000

25,000,000 = 25 × 1,000,000 = **25 million**

(this is a number-word combination)

Doing that "loses" 145,561, which is quite a bit! But sometimes it doesn't matter. This number represents the population of Texas from the 2010 census. To say that the Texas population in 2010 was about 25 million people is probably good enough for most situations (source: U.S. Census Bureau, http://quickfacts.census.gov/qfd/index.html).

Option 2: round 25,145,561 to the nearest hundred-thousand, which is 25,100,000.

25,100,000 = 25.1 × 1,000,000 = **25.1 million**

This option only "lost" 45,561.

Here is another example: 1,452,900,812 rounds to 1,500,000,000 or 1.5 billion

Resource **Order of Operations**

The order of operations defines the order in which operations are performed.

General Rule	Example
1) Operations within grouping symbols, innermost first. Grouping symbols include: • Parentheses ( ) • Brackets [ ] • Fraction Bar $\frac{\square}{\square}$	$15+[12-(3+2)]-2\times 3^2 \div 6$ $15+[12-(5)]-2\times 3^2 \div 6$ $15+[7]-2\times 3^2 \div 6$
2) Exponents	$15+[7]-2\times 9\div 6$
3) Multiplication and division, left to right	$15+[7]-2\times 18\div 6$ $15+[7]-3$
4) Addition and subtraction, left to right	$22-3$ 19

Resource **Probability, Chance, Likelihood, and Odds**

Probability

In Lesson 8, we use the word **risk** when talking about how likely it is that someone will get a disease. There are many other words that are used to describe this type of data. In mathematics, this is called a **probability**. The formula for calculating a probability is shown below. Note that you used the same type of division to calculate your percentages ("risk") in the lesson.

$$\text{Probability of an event} = \frac{\text{Number of times the event occurs}}{\text{Number of times the event could occur}}$$

If we call our event "E" and use "P" to represent probability, you sometimes will see this written as:

$$P(E) = \text{which means "the probability that E occurs"}$$

The parentheses are used differently here than their usual use as grouping symbols in mathematics; we could always use the complete phrase "the probability that E occurs." The "P(E)" notation is a shorthand for that phrase.

For example, we calculated the probability of any woman being diagnosed with lung cancer in a year:

$$\frac{\text{Number of women who get lung cancer}}{\text{Number of women}} = \frac{110{,}110 \text{ women}}{116{,}289{,}249 \text{ women}} = .0009469$$

So, P(any woman getting lung cancer) = .0009469

We read this statement as "the probability of any woman getting lung cancer is . . ."
Do you think that this probability is the same for all women and for each woman during their lives? You're right—this probability is an overall probability, and the probability for a given woman changes over time.

Other words that are used in describing probability are **chance** and **likelihood**.

> You can look at probability as a slight change in perspective from our table in Lesson 8.A. The table showed us that 90 women in 100,000 will get lung cancer.
>
> Probability lets us look at an average woman, and ask, "What is the likelihood that she will get lung cancer?"
>
> (Note: This is a general statement. Bear in mind that an individual woman can take steps to reduce her chance of lung cancer.)

Odds

In the media, you often hear **probability** and **odds** used interchangeably. However, they are not the same thing! Odds are best stated as a ratio.

$$\text{Odds of an event} = \frac{\text{Number of times the event occurs}}{\text{Number of times the event does not occur}}$$

So, the odds of a woman getting lung cancer =

$$\frac{\text{Number of women who get lung cancer}}{\text{Number of women who didn't get it}} = \frac{110{,}110 \text{ women}}{116{,}179{,}139 \text{ women}} = .0009478:1$$

(Do you see where the denominator came from?) We usually write odds as a ratio, such as the .0009478 to 1 in this example. The odds of getting lung cancer are not very different from the probability of getting lung cancer. Let's look at a better example:

If we flip a coin, what is the probability of getting "heads"?

$$P(\text{head}) = \frac{\text{number of heads}}{\text{number of outcomes}} = \frac{1}{2}$$

But what are the *odds* of getting "heads"?

$$\text{Odds of heads} = \frac{\text{number of heads}}{\text{number of "not heads"}} = \frac{1}{1}$$

We say "the odds of getting heads are 1 to 1."

Now say you have a jar with 5 marbles and 3 are green. You draw one marble from the jar.

The probability of drawing a green marble is $\frac{3}{5}$ or 60%.

The odds of drawing a green marble are 3 to 2.

Probabilities must be between 0 and 1. However, odds can be greater than 1, as in the example with these marbles. Probabilities can be given as percents, decimals, or ratios; odds are best given as ratios.

Resource **Properties**

On these pages, several important mathematical rules and relationships are given that can help you perform calculations in different ways. The formal names for the rules are also given. You do not have to memorize these names for this course, but you may use them in other math classes. If you want more help with any of the rules, use the formal names to find resources on the Internet.

The role of variables

The mathematical rules are defined in terms of **variables**. The variables are symbols, usually letters that represent numbers. You use variables to show that the rule can apply to a lot of different numbers. This is called **generalizing** because it shows that a rule can be used **in general** and not just in specific cases. The rules will be shown using variables and then give an example that uses numbers.

While mathematical rules are very important, in this course, the authors emphasize reasoning over memorizing the names of rules. As you review the rules, try to make sense of the rules so that they will become a part of your thinking.

Caution:

Be sure you notice which operations can be used with each property!

Communicative property

The order of addition and multiplication can be changed. It is important to remember that the Commutative Property does not apply to subtraction and division.

General Rule	Example
$a + b = b + a$	$8+3=3+8$
$a \times b = b \times a$	$5\times6=6\times5$

Distributive property

General Rule	Example
$a\ (b + c) = a \times b + a \times c$ also shown as $a\ (b + c) = ab + ac$ **Note about subtraction**: Subtraction is related to addition. The Distributive Property is shown using addition, but it also works with subtraction: $8\ (5 - 1) = 8 \times 5 - 8 \times 1$ **Notation**: The operation of multiplication is shown in many ways. In the above example, we see the multiplication symbol (x). We also see a number or variable in front of the parenthesis with no other symbol. For example: $6(2) = 6 \times 2$ $a(b) = a \times b$ You will learn other symbols for multiplication later in the course.	$4\ (3 + 1) = 4 \times 3 + 4 \times 1$ To demonstrate that these two calculations are equivalent, each side is done separately. **Left side**: Using Order of Operations, the operation inside the parentheses is done first. $4\ (3 + 1)$ $4\ (4)$ 16 **Right side**: Using the Distributive Property, the multiplication is *distributed* over the addition. $4\ (3 + 1)$ $4 \times 3 + 4 \times 1$ Order of Operations tells you to multiply first. $12 + 4$ 16

Division

Division is the same as multiplication by the reciprocal. You get the reciprocal of a number when you write the number as a fraction and reverse the numerator (the top number) and the denominator (bottom number). Example: The reciprocal of $\frac{2}{3}$ is $\frac{3}{2}$.

General Rule	Example
$a \div b = a \times \frac{1}{b}$	$15 \div 5 = 15 \times \frac{1}{5}$
$a \div \frac{b}{c} = a \times \frac{c}{b}$	$10 \div \frac{3}{5} = 10 \times \frac{5}{3}$
Caution: $15 \times \frac{1}{5}$ is the same as $\frac{1}{5} \times 15$ but $15 \div 5$ is not the same as $5 \div 15$	

Resource **Ratios and Fractions**

Language of ratios and fractions

A **ratio** is a comparison of two numbers by division. Ratios can be written.

- In words, such as *the ratio of males to females in class is 2 to 4.*
  This means there are two males for every four females in the class.
  Sometimes this is shown with a colon *males:females* and *2:4*.
- As a fraction, $\frac{\text{males}}{\text{females}} = \frac{2}{4}$.

Notice that the males are in the top of the fraction (the numerator) because they were mentioned first in the comparison. The females are in the bottom of the fraction (the denominator) because they were mentioned second.

$\frac{\text{males}}{\text{females}} = \frac{2}{4}$ simplifies to $\frac{1}{2}$. Be careful!

This does not mean that half the class is male! It means that there is one male for every two females. What is the ratio of males to the entire class?

Think about the class as groups of 2 males and 4 females. This means that two out of every six people are male.

$\frac{\text{males}}{\text{total}} = \frac{2}{6} = \frac{1}{3}$ indicates that the class is $\frac{1}{3}$ male.

This example shows how important it is to include the context along with the math (Writing Principle #2).

What is the ratio of females to males? What fraction of the class is female? What percent of the class is female?

The ratio of females to males is $\frac{\text{females}}{\text{males}} = \frac{4}{2} = \frac{2}{1}$.

The ratio of females to total people is $\frac{\text{females}}{\text{total}} = \frac{4}{6} = \frac{2}{3} \approx 0.667 \approx 66.7\%$.

2/3 of the class is female, which is about 66.7%. More about percentages later.

Resource **Rounding and Estimation**

Another important skill you used in this lesson is rounding. You often round numbers when you are trying to make sense out of them or make comparisons and do not need exact numbers. In this lesson, you estimated the length of a line of one billion people by measuring the people in your class as a sample. Since this was an estimate, there was no need to keep exact values in your results.

In this course, you will talk about different types of *estimation*.

- **Educated guess:** One type of estimation might be called an "educated guess" about something that has not been measured exactly. In an upcoming lesson, you will use estimations of the world population. This quantity cannot be measured exactly—it would be impossible to count how many people live on Earth at any given time. Scientists can use good data and mathematical techniques to estimate the population, but it will always be an estimate.
- **Convenient estimation:** Sometimes estimations are used when it is inconvenient or not worthwhile to make an exact count. Imagine that you need to know how much paint to buy to paint the baseboard trim in your house. (The baseboard trim is the piece of wood that follows along the bottom of the walls.) You need to know the length of the baseboard. You could measure the length of each wall to the nearest 1/8 inch and carefully subtract the width of halls and doors. It would be much quicker and just as effective to measure to the nearest foot or half foot. If you were cutting a piece of baseboard to go along the floor, however, you would want an exact measurement.
- **Estimated calculation:** This usually involves rounding numbers to make calculations simpler. A future lesson will focus on estimating and calculating percentages. In this course, you will find that percentages are used in many contexts. One of the most important skills that you will develop is understanding and being comfortable working with percentages in a variety of situations.

Resource **Scientific Notation**

Numbers can be written in many different forms. Some examples are:

$50 = 5 \times 10$

$400 = 40 \times 10$ or 4×100 or 4×10^2

$7{,}000 = 700 \times 10$ or 70×100 or 7×1000 or 7×10^3

Likewise: 52 can be written as 5.2×10

473 can be written as 47.3×10 or 4.73×100 or 4.73×10^2

$7{,}549 = 754.9 \times 10$ or 75.49×100 or 7.549×1000 or 7.549×10^3

Look at the last form in each row. What do they have in common? *They all have a single non-zero digit in front of the decimal place*. This means that we have a number greater than or equal to 1 but less than 10, multiplied by a power of 10. This is **scientific notation.**

A number in ***scientific notation*** is written in the form:

$a \times 10^n$ where $1 \leq a < 10$; and n is an integer.

1 is included and 10 is not included

7.549×10^3 is in scientific notation (7 is a number between 1 and 10).

75.49×10^2 is not in scientific notation (because 75 is not between 1 and 10).

7.549×100^3 is not in scientific notation because 100^3 is not written as a power of 10.

Resource Slope

Calculating Slope

In Lesson 16.C, you developed the formula for slope. In a linear relationship containing points (x_1, y_1) and (x_2, y_2), slope can be found in the following manner:

$$slope = m = \frac{y_2 - y_1}{x_2 - x_1}$$

Keep in mind:

- Slope can be found using any two points on a line. They can be given in a problem description, presented in a table, or read from a graph of the line.
- The "y" values are always the dependent values in the problem, even if you choose a different variable to represent them.
- The "x" values are always the independent values in the problem, even if you choose a different variable to represent them.

The points in each of the following examples are presented in different ways. Review the example, then try your own.

1) Using two points to find slope.

Example:

$(3,-2)$ **and** $(4,7)$

$$slope = m = \frac{y_2 - y_1}{x_2 - x_1}$$

$$m = \frac{7-(-2)}{4-3} = \frac{9}{1} = 9$$

You try:

$(-3,-5)$ **and** $(4,16)$

2) Use values in a table to calculate slope.

Example:

Time (sec)	5	10	15	20
Distance (yd)	40	75	110	145

Choose any two points: $(5,40)$ and $(15,110)$

$$slope = m = \frac{y_2 - y_1}{x_2 - x_1} \qquad m = \frac{110\text{yd} - 40\text{yd}}{15\text{sec} - 5\text{sec}} = \frac{70\text{yd}}{10\text{sec}} = \frac{7\text{yd}}{1\text{sec}}$$

We interpret the slope with the statement: *"The distance changes seven yards every one second."*

You try:

Snacks	5	10	15	20
Cost ($)	15	30	45	60

3) Use values from a graph to calculate slope.

Example:

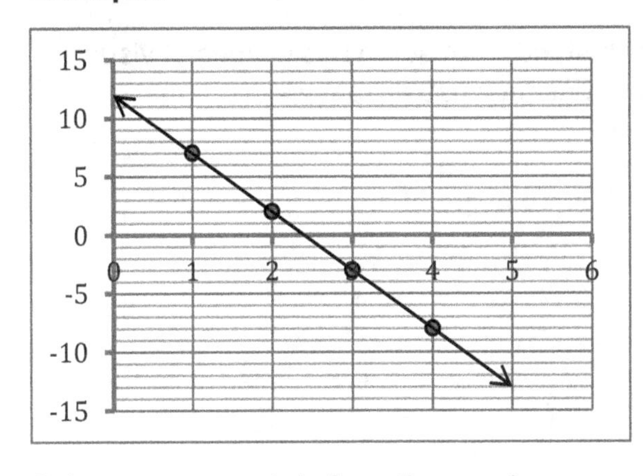

You try:

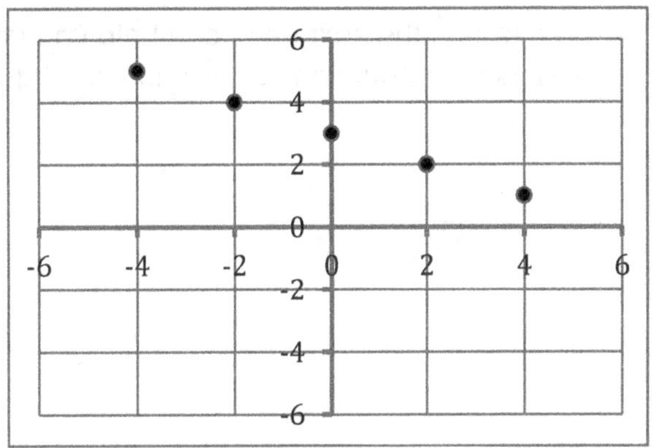

Select any two points from the graph.

$(2,2)$ and $(3,-3)$

$$slope = m = \frac{y_2 - y_1}{x_2 - x_1}$$

$$m = \frac{-3-2}{3-2} = \frac{-5}{1} = -5$$

Resource **Understanding Visual Displays of Information**

Asking questions about displays

Data are increasingly presented in a variety of forms intended to interest you and to invite you to think about the importance of these data and how they might affect your lives.

Some common displays are:

- pie charts
- scatterplots
- histograms and bar graphs
- line graphs
- tables
- pictographs

What questions should you ask yourself when you study a visual display of information?

- What is the title of the chart or graph?
- What question is the data supposed to answer? (For example: How many males versus females exercise daily?)
- How are the columns and rows labeled? How are the vertical and horizontal axes labeled?
- Select one number or data point and ask, "What does this mean?"

The chart on the next page can help you understand what some basic types of visual displays of information tell you and what questions they usually answer.

This looks like a ...	This visual display is usually used to ...	For example, it can be used to show ...
Pie chart	Show the relationships between different parts compared to a whole.	How time is used in a 24-hour cycle. How money is distributed. How something is divided up.
Line graph	Show trends over time. Compare trends of two different items or measurements.	What seems to be increasing. What is decreasing. How the cost of gas has increased in the last 10 years. Which of these foods (milk, steak, cookies, eggs) has risen most rapidly in price compared to the others.
Histogram or bar graph	Compare data in different categories. Show changes over time.	How a population is broken up into different age categories. How college tuition rates are changing over time.
Table	Organize data to make specific values easy to read. Break data up into overlapping categories.	The inflation rates over a period of years. How a population is broken into males and females of different age categories.

Resource **Writing Principles**

Writing background

You might be surprised that you are asked to write short responses to questions in this class. Writing in a math class? This course emphasizes writing for the following two reasons:

- Writing is a learning tool. Explaining things such as the meaning of data, how you calculated the data, or how you know your answer is correct deepens your own understanding of the material.
- Communication is an important skill in quantitative literacy. Quantitative information is used widely in today's world in products such as reports, news articles, publicity materials, advertising, and grant applications.

Understanding the task

One important strategy in writing is to make sure you understand the task. In this course, your tasks will be questions in assignments, but in other situations, the task might be a question on a report form, instructions from your employer, or a goal that you set for yourself. To begin to write successfully, ask yourself the following questions:

- What is the topic of the writing task?
- What is the task telling me to do? Some examples are given below:
  - Describe how you found the answer.
  - Explain why you think you have the right answer.
  - Reflect on the process of coming up with the answer.
  - Make a prediction about the next data point.
  - Compare two data points or the answers to two parts of the problem.
- What information am I given to help me with the task?

Look at this example and the answers to these questions.

> In Preview Assignment 2.A, you read about monitoring your readiness. Explain briefly why it is important to monitor your readiness before coming to class.

- What is the topic of the writing task? [Answer: Monitoring whether I am ready for the next class meeting.]
- What is the task telling me to do? [Answer: It is asking me to *explain* why "monitoring readiness" is important.]
- What information am I given to help me with the task? [Answer: I can look back at the Preview Assignment for 2.A if I need to refresh on this topic.]

Writing principles

Principle #1 If the problem has words, so should the answer!

(Have you noticed almost all of the problems in this class have words?)

Strive to be neat and to use proper grammar, spelling, and punctuation.

Principle #2 Each answer should be in a complete sentence that stands on its own, which means that the relevant information from the problem should be in the answer. **The readers should understand what you are trying to say even if they have not read the question or writing prompt.** Relevant information includes:

- Information about context.
- Quantitative information.

Refer back to question 8 from Student Page 2.A:

What are some factors you think may have led to this change in doubling times?

Insufficient response: *Improved health care, better food.*

Good response: *The world's population has increased rapidly. This increase may be due to factors such as improved health care, better food supplies, and clean water.*

Principle #3 If you use tables or graphs in your response, be sure they are clearly and thoroughly labeled.

Principle #4 Let the reader know if you are making any assumptions. One example of this could be when there is unclear information in the problem.

Mary's bedroom is 10 feet wide, 12 feet long, and 9 feet high. If a gallon of paint covers 300 square feet, how many gallons should Mary buy?

After doing the math and writing your answer, you might say, *"I didn't know how many windows and doors Mary has, so I just didn't take them into account in my calculations."*